I0049624

Improving Practice and Performance in Basketball

Improving Practice and Performance in Basketball

Special Issue Editors

Aaron T. Scanlan
Vincent J. Dalbo

MDPI • Basel • Beijing • Wuhan • Barcelona • Belgrade

MDPI

Special Issue Editors
Aaron T. Scanlan
Central Queensland University
Australia

Vincent J. Dalbo
Central Queensland University
Australia

Editorial Office
MDPI
St. Alban-Anlage 66
4052 Basel, Switzerland

This is a reprint of articles from the Special Issue published online in the open access journal *Sports* (ISSN 2075-4663) from 2017 to 2019 (available at: https://www.mdpi.com/journal/sports/special_issues/Improv_basketball).

For citation purposes, cite each article independently as indicated on the article page online and as indicated below:

LastName, A.A.; LastName, B.B.; LastName, C.C. Article Title. *Journal Name* **Year**, *Article Number*, Page Range.

ISBN 978-3-03921-694-9 (Pbk)
ISBN 978-3-03921-695-6 (PDF)

© 2019 by the authors. Articles in this book are Open Access and distributed under the Creative Commons Attribution (CC BY) license, which allows users to download, copy and build upon published articles, as long as the author and publisher are properly credited, which ensures maximum dissemination and a wider impact of our publications.

The book as a whole is distributed by MDPI under the terms and conditions of the Creative Commons license CC BY-NC-ND.

Contents

About the Special Issue Editors

Aaron T. Scanlan is a Senior Lecturer in Exercise and Sport Sciences and Director of the Human Exercise and Training Laboratory at Central Queensland University. Aaron received his Ph.D. undertaking applied sport science research in basketball at Central Queensland University in 2011. Since this time, Aaron has emerged as a world-leading researcher in basketball, with over 50 peer-reviewed journal publications focused specifically on basketball. Furthermore, Aaron is Chair of the Basketball Special Interest Group with the National Strength and Conditioning Association and has completed research projects working with international, national, and regional basketball teams. Aaron has a strong interest in improving the application of sport science in basketball practice, as well as the application of basketball to address societal issues.

Vincent J. Dalbo is a Senior Lecturer at Central Queensland University. Vincent received a B.S. in Psychology from University of Florida, a M.S. in Kinesiology from Georgia Southern University, and a Ph.D. in Exercise Physiology from the University of Oklahoma. Since 2007, Vincent has contributed to over 80 peer-reviewed publications and has been a part of teams that have received over $500,000 of research funding. Vincent's work has been conducted in the area of health and human performance and Vincent currently works as a sport scientist with several professional basketball teams and serves as Vice Chair of the Basketball Special Interest Group with the National Strength and Conditioning Association.

sports

MDPI

Editorial

Improving Practice and Performance in Basketball

Aaron T. Scanlan *[iD] and Vincent J. Dalbo[iD]

Human Exercise and Training Laboratory, School of Health, Medical and Applied Sciences,
Central Queensland University, Rockhampton 4702, Australia
* Correspondence: a.scanlan@cqu.edu.au; Tel.: +61-7-4923-2538

Received: 20 August 2019; Accepted: 21 August 2019; Published: 27 August 2019

Basketball is ranked in the top three team sports for participation in the Americas, Australia, Europe, Southeast Asia, and Western Pacific nations, making it one of the most popular team sports worldwide [1]. The physical demands and high popularity of basketball present a wide range of potential applications in society. At one end, basketball may offer a vehicle to combat high inactivity rates and reduce economic health burdens for government officials and health administrators in many countries due to the popularity of the game combined with the evidence supporting recreational basketball eliciting intense physical demands with low perceptual demand [2]. At the other end, professional basketball competitions have emerged in over 100 countries with more than 70,000 professional players globally [3], creating a lucrative business that provides legitimate career pathways for players and entertainment for billions of people. Despite the wide range in application, it is surprising how little research has been conducted in basketball relative to other sports. For instance, a rudimentary search on PubMed showed basketball to yield considerably less returns than other sports with a similar global reach and comparable returns to sports governed in less regions of the world (Table 1). Consequently, we sought to edit a Special Issue on "Improving Practice and Performance in Basketball" to provide a collection of studies from basketball researchers across the world and increase available evidence on pertinent topics in the sport. In total, 40 researchers from 16 institutions or professional bodies across nine countries contributed 10 studies in the Special Issue.

Table 1. Returns on Scopus for basketball relative to other sports.

Search Term (Sport)	Number of Returns	Number of Countries Played in *
Soccer	4176	211
Tennis	2188	211
Basketball	1427	213
Baseball OR Softball	1250	141
"American football" OR NFL	1150	104
Hockey	993	137 [†]
Cricket	940	104
"Rugby union" OR "Rugby sevens" OR "Rugby league"	808	119 [‡]

Note: Search conducted on August 9th 2019 via https://www.ncbi.nlm.nih.gov/pubmed/ and was restricted to past five years. * The number of countries identified as members by the international governing body; [†] field hockey included 137 member countries, while ice hockey included 76 member countries; [‡] rugby union included 119 member countries, while rugby league included 68 member countries.

Most research conducted in basketball has focused on athletic populations. For instance, a review of the 228 studies returned on PubMed for "basketball" in 2019 (up to August 9th) indicates over 25% of studies focused on the incidence, treatment, rehabilitation, or screening of injuries, while 11% of studies described physical, fitness, or functional attributes in competitive basketball players. These trends emphasize the strong interest in understanding injury prevention and treatment in basketball, as well as attributes which may underpin successful players, both of which are oriented towards optimizing player and team performance. Regarding enhancing performance, an increasingly popular field of research in basketball is examining monitoring methods (7% of PubMed studies in 2019) to better understand demands placed on players across the season and provide evidence for decision-making regarding player management. Several reviews have recently been published highlighting the interest in quantifying game [4] and training demands [5], using heart rate monitoring [6], and applying microsensors to measure player workloads [7] in competitive basketball. Available monitoring technologies provide basketball coaches and high-performance staff with a plethora of data regarding player fitness, workloads, and fatigue status to inform decisions regarding training prescription and recovery opportunities for minimizing injury risk and optimizing performance. In turn, basketball research has readily used game-related statistics (3% of PubMed studies in 2019) to describe player and team performance, which provide an expansive reservoir of data, usually publicly available, to link outcomes of interest to performance. Consequently, our Special Issue was open to research exploring various current topics that have potential to impact practice in basketball.

In keeping with the recent trends in basketball research, the Special Issue contains two reviews with one focused on exploring the utility of various monitoring strategies to detect player fatigue [8] and the other identifying issues to consider around the extensive travelling requirements in the National Basketball Association (NBA), the premier global basketball competition [9]. Both reviews highlight the practical aspects relating fatigue and travel in basketball, including potential implications for injury, workload management, recovery, and assessment in players. Furthermore, two applied studies in the Special Issue examine workload monitoring in basketball, with one exploring the impact of game scheduling on accelerometer-derived workload [10] and the other examining changes in jump kinetics and perceptual workload across the season [11]. An additional three studies in the Special Issue identified game-related statistics explaining game outcomes and regional differences in various elite competitions (Olympics [12], EuroBasket [13], and Continental Championships [14]). The remaining three studies described physical [15,16] and skill [17] attributes in various player samples. It should also be noted our Special Issue addresses an important issue of increasing research in female athletes, who have traditionally been under-represented in the literature compared to male basketball players, with seven of the eight original studies (88%) containing female basketball players.

The immediate future of basketball research in high-performance settings is highlighted by issues faced in practice. Specifically, key players are missing games or being rested for "load management" in the NBA to reduce player injury risk, despite some initial evidence suggesting greater rest during the regular season (6 ± 1 vs 1 ± 1 games) does not reduce injury incidence or performance in the playoffs [18]. Likewise, condensed game schedules [19] and the total minutes played in individual games [20] have been shown to have no significant effects on injury risk in NBA players. In contrast, other research suggests the total number of games played in a season impacts injury risk in the NBA [21], highlighting the need for further research on this topic to gather a definitive understanding regarding the effects of managing player workloads on injury risk. In fact, more research needs to build upon the extensive descriptive evidence already available and identify modifiable factors contributing to injuries in basketball players for coaches and high-performance staff to control risk as much as possible. In addition to injury, future basketball research should seek to further examine the efficacy of logical and practical intervention strategies on player performance. For example, an increasing number of studies are examining the utility of different training approaches, including resistance training [22], court-based conditioning [23], and games-based drills [5], as well as nutritional strategies [24–27] and recovery practices [28] on performance outcomes. Furthermore, it is integral for future research assessing player performance to use basketball-specific assessments. In this regard, more research is recognizing the need for greater specificity in measuring performance in basketball, with an increased number of studies exploring the utility of basketball-specific testing protocols to assess relevant physical attributes [29–32] as well as in-game statistics [33] and workloads [34] to quantify player performance in a robust manner with increased application to actual competition.

Sports **2019**, *7*, 197

Conflicts of Interest: The authors declare no conflict of interest.

References

1. Hulteen, R.; Smith, J.; Morgan, P.; Barnett, L.; Hallal, P.; Colyvas, K.; Lubans, D. Global participation in sport and leisure-time physical activities: A systematic review and meta-analysis. *Prev. Med.* **2017**, *95*, 14–25. [CrossRef] [PubMed]
2. Stojanovic, E.; Stojiljkovic, N.; Stankovic, R.; Scanlan, A.; Dalbo, V.; Milanovic, Z. Recreational basketball small-sided games elicit high-intensity exercise with low perceptual demand. *J. Strength Cond. Res.* **2019**, in press. [CrossRef] [PubMed]
3. Basketball Leagues Around the World. Available online: https://jazzteamstore.com/blog/basketball-leagues-around-the-world/ (accessed on 9 August 2019).
4. Stojanović, E.; Stojiljkovic, N.; Scanlan, A.; Dalob, V.; Berkelmans, D.; Milanovic, Z. The activity demands and physiological responses encountered during basketball match-play: A systematic review. *Sports Med.* **2018**, *48*, 111–135. [CrossRef] [PubMed]
5. Clemente, F. Small-sided and conditioned games in basketball training: A review. *Strength Cond. J.* **2016**, *38*, 49–58. [CrossRef]
6. Berkelmans, D.; Dalbo, V.; Kean, C.; Milanovic, Z.; Stojanovic, E.; Stojiljkovic, N.; Scanlan, A. Heart rate monitoring in basketball: Applications, player responses, and practical recommendations. *J. Strength Cond. Res.* **2018**, *32*, 2383–2399. [CrossRef] [PubMed]
7. Fox, J.; Scanlan, A.; Stanton, R. A review of player monitoring approaches in basketball: Current trends and future directions. *J. Strength Cond. Res.* **2017**, *31*, 2021–2029. [CrossRef] [PubMed]
8. Edwards, T.; Spiteri, T.; Piggott, B.; Bonhotal, J.; Haff, G.; Joyce, C. Monitoring and managing fatigue in basketball. *Sports* **2018**, *6*, 19. [CrossRef]
9. Huyghe, T.; Scanlan, A.; Dalbo, V.; Calleja-Gonzalez, J. The negative influence of air travel on health and performance in the National Basketball Association: A narrative review. *Sports* **2018**, *6*, 89. [CrossRef]
10. Staunton, C.; Wundersitz, D.; Gordon, B.; Cutovic, E.; Stanger, J.; Kingsley, M. The effect of match schedule on accelerometry-derived exercise dose during training sessions throughout a competitive basketball season. *Sports* **2018**, *6*, 69. [CrossRef]
11. Legg, J.; Pyne, D.; Semple, S.; Ball, N. Variability of jump kinetics related to training load in elite female basketball. *Sports* **2017**, *5*, 85. [CrossRef]
12. Leicht, A.; Gomez, M.; Woods, C. Team performance indicators explain outcome during women's basketball matches at the Olympic games. *Sports* **2017**, *5*, 96. [CrossRef]
13. Conte, D.; Lukonaitiene, I. Scoring strategies differentiating between winning and losing teams during FIBA EuroBasket women. *Sports* **2017**, *6*, 50. [CrossRef]
14. Madarame, H. Regional differences in women's basketball: A comparison among Continental Championships. *Sports* **2018**, *6*, 65. [CrossRef]
15. Gryko, K.; Kopiczko, A.; Mikolajec, K.; Stasny, P.; Musalek, M. Anthropometric variables and somatotype of young and professional male basketball players. *Sports* **2018**, *6*, 9. [CrossRef]
16. Fields, J.; Merrigan, J.; White, J.; Jones, M. Seasonal and longitudinal changes in body composition by sport-position in NCAA Division I basketball athletes. *Sports* **2018**, *6*, 85. [CrossRef]
17. Policastro, F.; Accardo, A.; Marcovich, R.; Pelamatti, G.; Zoia, S. Relation between motor and cognitive skills in Italian basketball players aged between 7 and 10 years old. *Sports* **2018**, *6*, 80. [CrossRef]
18. Belk, J.; Marshall, H.; McCarty, E.; Kraeutler, M. The effect of regular-season rest on playoff performance among players in the National Basketball Association. *Orthop. J. Sports Med.* **2017**, *5*, 2325967117729798. [CrossRef]
19. Teramoto, M.; Cross, C.; Cushman, D.; Maak, T.; Petron, D.; Willick, S. Game injuries in relation to game schedules in the National Basketball Association. *J. Sci. Med. Sport* **2017**, *20*, 230–235. [CrossRef]
20. Okoroha, K.; Marfo, K.; Meta, F.; Matar, R.; Shehab, R.; Thompson, T.; Moutzouros, V.; Makhni, E. Amount of minutes played does not contribute to anterior cruciate ligament injury in National Basketball Association athletes. *Orthopedics* **2017**, *4*, e658–e662. [CrossRef]

21. Talukder, H.; Vincent, T.; Foster, G.; Hu, C.; Huerta, J.; Kumar, A.; Malazarte, M.; Saldana, D.; Simpson, S. *Preventing in-Game Injuries for NBA Players*; MIT Sloan Sports Analytics Conference: Boston, MA, USA, 2016; p. 1590.

22. Freitas, T.; Calleja-Gonzalez, J.; Carlos-Vivas, J.; Marin-Cascales, E.; Alcaraz, P. Short-term optimal load training vs a modified complex training in semi-professional basketball players. *J. Sports Sci.* **2019**, *37*, 434–442. [CrossRef]

23. Pliauga, V.; Lukonaitiene, I.; Kamandulis, S.; Skurvydas, A.; Sakalauskas, R.; Scanlan, A.; Stanislovaitiene, J.; Conte, D. The effect of block and traditional periodization training models on jump and sprint performance in collegiate basketball players. *Biol. Sport* **2018**, *35*, 373–382. [CrossRef] [PubMed]

24. Afman, G.; Garside, R.; Dinan, N.; Gant, N.; Gant, N.; Betts, J.; Williams, C. Effect of carbohydrate or sodium bicarbonate ingestion on performance during a validated basketball simulation test. *Int. J. Sport Nutr. Exerc. Metab.* **2014**, *24*, 632–644. [CrossRef] [PubMed]

25. Delextrat, A.; Mackessy, S.; Arceo-Rendon, L.; Scanlan, A.; Ramsbottom, R.; Calleja-Gonzalez, J. Effects of three-day serial sodium bicarbonate loading on performance and physiological parameters during a simulated basketball test in female university players. *Int. J. Sport Nutr. Exerc. Metab.* **2018**, *25*, 547–552. [CrossRef]

26. Scanlan, A.; Dalbo, V.; Conte, D.; Stojanovic, E.; Stojiljkovic, N.; Stankovic, R.; Antic, V.; Milanovic, Z. Caffeine supplementation has no effect on dribbling speed in elite basketball players. *Int. J. Sports Physiol. Perform.* **2019**, *14*, 997–1000. [CrossRef] [PubMed]

27. Stojanovic, E.; Stojiljkovic, N.; Scanlan, A.; Dalbo, V.; Stankovic, R.; Antic, V.; Milanovic, Z. Acute caffeine supplementation promotes small to moderate improvements in performance tests indicative of in-game success in professional female basketball players. *Appl. Physiol. Nutr. Metab.* **2019**, *44*, 849–856. [CrossRef] [PubMed]

28. Delextrat, A.; Hippocrate, A.; Leddington-Wright, S.; Clarke, N. Including stretches to a massage routine improves recovery from official matches in basketball players. *J. Strength Cond. Res.* **2014**, *28*, 716–727. [CrossRef]

29. Scanlan, A.; Wen, N.; Pyne, D.; Stojanovic, E.; Milanovic, Z.; Conte, D.; Vaquera, A.; Dalbo, V. Power-related determinants of Modified Agility T-test Performance in male adolescent basketball players. *J. Strength Cond. Res.* **2019**, in press. [CrossRef] [PubMed]

30. Boddington, B.; Cripss, A.; Scanlan, A.; Spiteri, T. The validity and reliability of the Basketball Jump Shooting Accuracy Test. *J. Sports Sci.* **2019**, *37*, 1648–1654. [CrossRef]

31. Wen, N.; Dalbo, V.; Burgos, B.; Pyne, D.; Scanlan, A. Power testing in basketball: Current practice and future recommendations. *J. Strength Cond. Res.* **2018**, *32*, 2677–2691. [CrossRef]

32. Scanlan, A.; Fox, J.; Borges, N.; Delextrat, A.; Spiteri, T.; Dalbo, V.; Stanton, R.; Kean, C. Decrements in knee extensor and flexor strength are associated with performance fatigue during simulated basketball game-play in adolescent, male players. *J. Sports Sci.* **2018**, *36*, 852–860. [CrossRef]

33. Staunton, C.; Gordon, B.; Custovic, E.; Stanger, J.; Kingsley, M. Sleep patterns and match performance in elite Australian basketball athletes. *J. Sci. Med. Sport* **2017**, *20*, 786–789. [CrossRef] [PubMed]

34. Fox, J.; Stanton, R.; Scanlan, A. A comparison of training and competition demands in semiprofessional male basketball players. *Res. Q. Exerc. Sport* **2018**, *89*, 103–111. [CrossRef] [PubMed]

© 2019 by the authors. Licensee MDPI, Basel, Switzerland. This article is an open access article distributed under the terms and conditions of the Creative Commons Attribution (CC BY) license (http://creativecommons.org/licenses/by/4.0/).

sports

MDPI

Review

Monitoring and Managing Fatigue in Basketball

Toby Edwards [1,*], Tania Spiteri [1,2], Benjamin Piggott [1], Joshua Bonhotal [3], G. Gregory Haff [2] and Christopher Joyce [1]

[1] School of Health Sciences, The University of Notre Dame Australia, Fremantle, WA 6959, Australia; tania.spiteri@nd.edu.au (T.S.); benjamin.piggott@nd.edu.au (B.P.); chris.joyce@nd.edu.au (C.J.)
[2] Centre for Exercise and Sport Science Research, Edith Cowan University, Perth, WA 6027, Australia; g.haff@ecu.edu.au
[3] Purdue University, Purdue Sports Performance, West Lafayette, IN 47906, USA; jbonhotal@purdue.edu
[*] Correspondence: tobyedwards5@gmail.com; Tel.: +1-765-490-0736

Received: 8 January 2018; Accepted: 22 February 2018; Published: 27 February 2018

Abstract: The sport of basketball exposes athletes to frequent high intensity movements including sprinting, jumping, accelerations, decelerations and changes of direction during training and competition which can lead to acute and accumulated chronic fatigue. Fatigue may affect the ability of the athlete to perform over the course of a lengthy season. The ability of practitioners to quantify the workload and subsequent fatigue in basketball athletes in order to monitor and manage fatigue levels may be beneficial in maintaining high levels of performance and preventing unfavorable physical and physiological training adaptations. There is currently limited research quantifying training or competition workload outside of time motion analysis in basketball. In addition, systematic research investigating methods to monitor and manage athlete fatigue in basketball throughout a season is scarce. To effectively optimize and maintain peak training and playing performance throughout a basketball season, potential workload and fatigue monitoring strategies need to be discussed.

Keywords: microtechnology; smallest worthwhile change; training load; countermovement jump

1. Introduction

Basketball is an intermittent, court-based team sport comprised of repeated high intensity movements such as change of direction, accelerations and decelerations interspersed with periods of low to moderate intensity activity [1]. Athletes also perform regular maximal efforts during competition including extensive high intensity shuffling, sprinting and jumping [2,3]. Research using time motion analysis (TMA) to investigate the competition demands of basketball have revealed that the mean distance covered by female and male athletes was 5–6 km during live playing time across 40 min games [1]. Physiological traits such as blood lactate and heart rate responses to competition demands reveal athletes are competing at an average physiological intensity above lactate threshold and 85% maximum heart rate [1]. The competition demands encountered by basketball players suggest that both anaerobic and aerobic energy pathways contribute to energy sources. Basketball also has one of the longest seasons in professional sports. Typically, a professional National Basketball Association (NBA) season consists of 82 games played over six months. If successful, teams can play over 100 games if they make post season play offs. Competitive seasons in Division I collegiate basketball in the United States span five months and include approximately 30 regular season games which is consistent with other semi-professional and professional leagues around the world. The high intensity movement demands and physiological stress on the athletes during competition may accumulate over the pre-season and competitive season and present as signs of fatigue leading to decreased performance output and/or injury [4]. Combining objective and subjective measures of workload and fatigue provides practitioners such as strength and conditioning coaches and sport scientists with a global picture of how the athlete is responding to the training dose, competition demands and non-training stressors. Early identification and subsequent management of fatigue may prevent detrimental physical and physiological adaptations often associated with injury and enhance athletic performance and player availability [5,6].

It is well understood that fatigue can inhibit athletic performance [4,7–10], however, conflicting definitions of fatigue make monitoring and measuring the underlying fatigue mechanisms problematic. The two attributes of fatigue that need to be acknowledged are: perceived fatigability, the maintenance of homeostasis and subjective psychological state of the athlete; and performance fatigability, the decline in objective performance measures derived from the capacity of the nervous system and contractile properties of muscles over time [11] (Figure 1). To align with a recent report [11], this review will define fatigue as a symptom where cognitive and physical function is limited by the interaction between perceived fatigability and performance fatigability. These two measures are able to normalize observed fatigue to the demands associated with the sport. For example, athletes who are less fatigable are able to endure a greater amount of workload before reaching a given level of fatigue [11]. Whilst multiple methods to monitor and manage perceived and performance fatigue in other sports have been investigated such as sprint speed [12,13], wellness questionnaires [14–16], biochemical markers [17] and neuromuscular tests [18], there is a lack of research examining the longitudinal use of these methods and practices in basketball. This review aims to provide practitioners with an overview of fatigue monitoring tools and management methods that have been reported in the literature or appear suitable in collegiate or professional basketball.

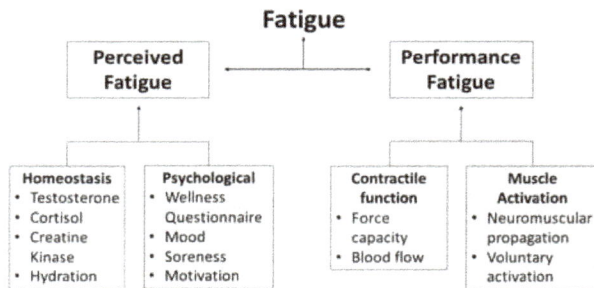

Figure 1. Modulating factors of perceived and performance fatigability (Adapted from [11]).

2. Materials and Methods

The search strategy used to locate articles included an online search of journal databases including PubMed, Web of Science, EBSCO host and Google Scholar. Key terms used in the search included monitoring OR managing AND fatigue OR performance AND basketball. In addition, articles cited in the reference lists of identified journals were manually searched and examined.

3. Quantifying Workload

Before managing fatigue, it is important to quantify and understand the training and competition workload the athlete has completed. Combining the athletes' workload and fatigue measurement will allow practitioners to determine the dose-response relationship and help inform whether the athlete is prepared for competition. Since the physical and physiological adaptations from a training stimulus vary between individuals based on modifiable (strength, aerobic/anaerobic capacity) and non-modifiable (age, gender, anatomy, genetics) factors, it is necessary to monitor the individual dose-response relationship [6]. For example, a strong correlation was detected in elite European basketball athletes between distance covered in the Yo-Yo intermittent recovery level one test and session rating of perceived exertion (s-RPE) scores during practice (r = 0.68) [19]. This suggests that assuming athletes achieved equal amount of workload, athletes with an increased aerobic capacity perceive the same training session as being easier than athletes who have a lower aerobic capacity. Currently, TMA and s-RPE are the most common methods to quantify the movement and workload demands in basketball [20], however recent advances in technology have allowed microtechnology devices to objectively quantify the external load of athletes in training and competition. A recent review of player monitoring approaches in basketball extensively discusses the advantages and disadvantages of several methods to quantify an athlete's workload [20]. A unique aspect of this review is that it adds to Fox and colleagues [20] review by briefly discussing the findings and results of previous research that has investigated basketball training and/or competition demands using microtechnology or s-RPE.

3.1. Microtechnology

Microtechnology has become a popular tool for practitioners and researchers to monitor and quantify the physical demands of athletes during training and competition in outdoor field sports such as soccer, rugby league, rugby union and Australian Rules football (AF) [21,22]. However, quantifying the external demands of basketball using microtechnology is challenging due to several limitations including that the game is played in indoor stadiums, the feasibility of acquiring enough units and the reliability and validity of microtechnology to detect basketball specific movements [22]. Recently, advances in technology have integrated a number of micro inertial sensors including triaxial accelerometers, gyroscopes and magnetometers into single units commonly referred to as inertial measurement units (IMU) [22]. These devices have assisted in overcoming some of the previously mentioned limitations that surround quantifying movement demands in basketball training and competition. The IMU provides an array of information that can inform practitioners' decisions on the performance of basketball players in training and competition including the position, direction, velocity, accelerations and decelerations [21,22]. A recent study used a tri-axial accelerometer with a sample rate of 100 Hz to examine the external demands of common training drills [23]. Instantaneous data from all 3 axes (x, y and z) were assimilated into a resultant vector through the Cartesian formula $\sqrt{[(x_n - x_{n-1})^2 + (y_n - y_{n-1})^2 + (z_n - z_{n-1})^2]}$. Accelerometer Load (AL) for each drill and activity was then calculated by summing the instantaneous change of rates of resultant accelerations over time [23]. The authors reported full court 3v3 and 5v5 (18.7 ± 4.1; 17.9 ± 4.6 AL/min, respectively) produced greater AL than full court 2v2 and 4v4 (14.6 ± 2.8; 13.8 ± 2.5 AL/min, respectively) [23]. In regards to playing position, the authors reported higher AL for point guards irrespective of training drill. This may represent tactical requirements of the position as smaller players may be required to cover more distance per possession. Another logical reason for guards to have greater AL is that they are able to accelerate easier with less applied force due to lower body mass [23]. A more recent investigation into training and competition demands of semi-professional basketball players reported significantly higher absolute and relative AL during game based training than competition (624 ± 113 AL vs. 449 ± 118 AL, ES = 1.54; 6.10 ± 0.77 AL/min vs. 4.35 ± 1.09 AL/min, ES = 2.14 respectively) [24]. This shows that pre-season training in semi-professional basketball appears to adequately prepare players for competition. The combination of these findings, and the application of IMUs in basketball, may help practitioners improve athletes' conditioning by developing position specific drills, improve training periodization, and provide more accurate drill clarification and description. However, systematic monitoring of external demands using IMUs is still warranted to provide a greater understanding of the suitability and effectiveness of the devices in basketball to quantify basketball activities such as shuffling and jumping.

3.2. Session Rating of Perceived Exertion

An issue that practitioners are commonly faced with when quantifying an athlete's workload is that the different scale, units and type vary across different training modalities. For example, comparing the load of a resistance training session (sets × reps × weight) and a court based training session (accelerations, decelerations, velocity) is problematic as there is no single objective load monitoring variable for both modalities of training. The simple method of s-RPE can overcome this issue and be used across several training modalities to monitor an athlete's perceived exertion from a particular training session [25,26] and longitudinally across an entire season [27,28]. By using a modified Borg RPE scale ranging from 0 to 10 which represent rest and maximal exertion respectively, athletes can provide a subjective rating of the intensity of a particular training session [26]. This number is multiplied by the duration in minutes to provide an arbitrary unit of subjective internal training load [26]. Unlike microtechnology, s-RPE has been widely reported in basketball literature that has investigated internal responses to training and competition [3,19,29–32]. In elite European basketball weekly s-RPE training load significantly differed between the control week (no game) and those accumulated during 1 or 2 game week microcycles (3334 ± 256 vs. 2928 ± 303 vs. 2791 ± 239 arbitrary units (AU), respectively) [19]. In addition, authors reported a strong correlation between s-RPE and heart rate based training load model (Edwards' TRIMP) in the same population (r = 0.68) [19]. However, s-RPE exhibited a moderate relationship (r = 0.49) and low commonality (R^2 = 0.24) with accelerometer derived training load in semi-professional Australian basketball players [32]. This suggests that s-RPE measures different training constructs than external AL. Therefore, it is recommended that practitioners collect both external and internal training load measures such as s-RPE and accelerometer or IMU training load as the intermittent demands and lateral movements required in basketball can increase an athlete's s-RPE by 13–25% when external load is controlled [33]. S-RPE is non-invasive, simple to calculate and quantify across the length of a basketball season making it an efficient and practical tool to use in both research and practice.

4. Fatigue Monitoring Tools

A number of different fatigue monitoring tools exist that may assist practitioners in identifying indicators of performance and perceived fatigability in basketball athletes including sprinting ability, vertical jumps, athlete self-report measures (ASRM), heart rate indices and biochemical markers. These fatigue monitoring tools may be beneficial in monitoring athletes' fatigue levels during a long season where accumulation of fatigue may affect player on court performance. Incorporating several fatigue monitoring tools simultaneously may provide practitioners with a global understanding of how athletes are responding to training and non-training stressors. Subsequently, a player's prescribed workload can be altered as necessary.

4.1. Sprinting Ability

Sprinting is a critical movement performed by all players during basketball training and competition [34]. Sprint speed has been identified as an important attribute of basketball athletes, specifically 5 m sprint times has exhibited a moderate inverse relationship to playing time (r = −0.59) in the NCAA Division II competition [35]. Conversely, 20 m sprint time demonstrated a weak correlation to total playing time [35,36] and basketball specific statistics including points, assists, rebounds, steals and blocks [37]. Monitoring an athlete's acceleration ability may be a more appropriate method to identify fatigue in basketball athletes in contrast to maximal speed as players rarely sprint the length of the court and therefore do not reach maximal speed in competition [34]. However, conjecture surrounds the use of sprint assessments as fatigue monitoring tools in previous literature. In rugby league, non-significant changes in 10 and 40 m were reported following six weeks of deliberate overreaching [28]. Meanwhile, in soccer players, 20 yard (18.3 m) sprint times decreased in starters and not in non-starters during 11 weeks of soccer competition [13]. Specifically, to basketball 10 m sprint time was decreased up until 24 h post-match (ES = 0.5) in elite European basketball players [38]. Therefore, monitoring acceleration ability over 5 to 10 m seems promising as a measure of performance fatigability in basketball.

4.2. Athlete Self-Report Measures

A recent survey on fatigue monitoring tools in high performance sport reported a high usage of ASRM across various sports and levels of competition for assessing overall well-being of team sport athletes [4]. Several ASRM have been used in the literature including the Profile of Athlete Mood States (POMS) [39], Daily Analysis of Life Demands of Athletes (DALDA) [40], Total Quality Recovery (TQR) [41] and the Recovery Stress Questionnaire for Athletes (REST-Q) [28]. However, to minimize time constraints on athletes, many team sport practitioners prefer shorter, customized versions that can be completed on a daily basis [4]. The shorter customized ASRM has been shown to be sensitive to daily, weekly and seasonal changes in training load in elite AF and English soccer players [14,16,42]. Specifically, daily ASRM that included fatigue, sleep quality, stress, mood and muscle soreness significantly associated with daily fluctuations in training load during the pre-season and competitive periods of elite AF and English soccer players respectively [16,42]. More recently, pre-training subjective ASRM have been suggested to provide practitioners with information on an athlete's capacity to train [7,15]. For example, in American collegiate football an increase of one unit in muscle soreness (players felt less sore) z score led to a trivial 4.4% decrease in s-RPE training load [7]. In AF, a one unit decrease in wellness z score corresponded to 4.9% decrease in player load [15]. The z score indicates how many standard deviations a variable is from the mean and can be calculated using the following formula: z score = athlete's score—athlete's mean score/standard deviation (SD) of athlete's score [7,8,15]. Whilst there is limited research investigating customized ASRM in basketball, the evidence in several other sports suggest that lower pre-training wellness scores may lead to a decrease in external load and an increase in internal load [7,9,15,40]. Implementing daily ASRM into an athlete monitoring program for basketball athletes may assist practitioners in understanding the perceptual fatigue of athletes, how they are coping with training and competition schedule, and also provide insight into intensity of output expected from an athlete in training.

4.3. Vertical Jumps

The use of vertical jump performance as a fatigue monitoring tool is also popular in high performance sport to assess lower body strength and power, and the integrity of the musculotendinous pre-stretch, or countermovement stretch shortening cycle (SSC) [43,44]. More than half of the respondents (54%) in a fatigue monitoring survey reported using vertical jump testing on either a daily, weekly or monthly basis to monitor performance and neuromuscular fatigue [4]. A variety of offensive and defensive movements are completed by basketball athletes during training and competition including accelerating, decelerating and change of direction that rely heavily on the athletes ability to rapidly transition from eccentric to concentric contraction via the SSC [45]. Repetitive performance of these movements can result in reduced movement efficiency through neuromuscular and performance fatigue [43,44]. Several vertical jump protocols have been used to monitor neuromuscular function and the SSC including the drop jump (DJ) and the countermovement jump (CMJ) [4,9]. In addition, a number of different apparatus have also been used in the literature to monitor vertical jump performance in athletes including a Vertec system (jump and reach) [12], contact mats [46], force plates [18] and linear position transducers [45]. Many of these protocols and instruments can be administered, analyzed and reported quickly in order to make decisions regarding the athlete's daily or weekly training prescription.

A meta-analysis reported using the average height of multiple CMJs was more sensitive in detecting CMJ fatigue and supercompensation than the maximum CMJ height [47]. However, conflicting evidence surrounds the use of jump height as the sole fatigue monitoring variable. For example, results from a 3 day elite handball competition demonstrated significant decline in CMJ height [48] though no changes in CMJ height were observed in elite rugby sevens players during the final preparation period [49]. The inconsistent findings surrounding jump height as a global indicator of neuromuscular function and performance fatigue is likely due to its gross representation of several underlying kinematic variables that contribute to CMJ height. These underlying variables that contribute to CMJ height relate to the eccentric and/or concentric phase and may provide a greater insight into the integrity of the SSC, loading strategies and behaviors used to execute a CMJ [50]. Findings from a study investigating the response of a CMJ following training and competition suggest flight time to contraction time (FT:CT) ratio appears to be a sensitive measure able to detect neuromuscular fatigue in female basketball athletes [51]. In contrast, basketball players reactive strength index (flight time/contact time) derived from a 40 cm drop jump was not sensitive to detect changes in s-RPE training load during a competitive elite Australian basketball season [46]. It is difficult to make comparisons between the two findings as they both elicit different loading strategies and behaviors. For example, the CMJ assesses a slow SSC response (contact time >250 ms) whilst the DJ elicits a fast SSC response (contact time <250 ms) [46]. Despite this, administering vertical jump performance test as fatigue monitoring tools seem promising as high levels of neuromuscular function is critical to vertical jumping capacity, change of direction ability and basketball performance [52].

4.4. Heart Rate

The autonomic nervous system (ANS) is linked with many physiological systems and can potentially identify fatigue and negative training adaptations through alterations in heart rate [53]. Specifically, several heart rate derived metrics including resting heart rate (RHR) and heart rate variability (HRV) have the potential to provide practitioners with an understanding of how an athlete is responding to fluctuations in training and competition workload. The use of heart rate metrics for monitoring athlete fatigue has been comprehensively reviewed [53], therefore the following will provide a brief overview of each variable and the applicability to basketball.

4.4.1. Resting Heart Rate

One of the first signs of overtraining syndrome commonly reported in the literature is an increase in RHR [54]. However conflicting research exists with some early investigations reporting increased RHR in overreaching athletes and those with overtraining syndrome [55], whilst other studies found RHR remained similar in overreaching and normal states [56,57]. A systematic review of 34 studies investigated whether RHR can be used to determine overreaching in athletes reported moderate increase in RHR after short (<2 weeks) interventions but no difference was found in longer (>2 weeks) interventions [54]. These findings suggest that the use of RHR to monitor fatigue in basketball athletes may be beneficial during intensive training camps (<2 weeks) and congested fixtures where spikes in workload are common potentially leading to an increase in fatigue. Consequently, including RHR in a longitudinal athlete monitoring system over the length of a season or to monitor non-functional overreaching or over training syndrome may not provide a valid sign of fatigue.

4.4.2. Heart Rate Variability

Research investigating changes in HRV in athletes during heavy training and competition periods has received increased interest due to the high reliability and the ability to capture data over a short period (~60 s) [58,59]. A common interval period often used as an index of ANS responsiveness is known as the R-R interval, or the time between heart beats. Whilst RHR can remain relatively stable, vagal related time periods can vary substantially [59]. However, conflicting findings are reported in the literature in relation to the use of HRV as a fatigue monitoring tool. Specifically, HRV has demonstrated sensitivity to changes in workload and performance in individual sports such as weightlifting [60], swimming [61] and middle-distance running [62] with only trivial evidence in team sport athletes [9,10]. In spite of the support of HRV in individual athletes, a systematic review reported only small effects of overreaching on HRV [54]. Similarly to RHR, this finding was also limited to short (2 weeks) interventions/overload [54]. An absence of research in determining the use of HRV in basketball athletes suggest that more research is needed to further clarify its usefulness as a fatigue monitoring tool. In addition, previous research in team sports suggest using caution if including HRV in an athlete monitoring program [9,10,54].

4.5. Biochemical Markers

When prescribing an athlete's workload to optimize training adaptations and avoid inducing further fatigue, it is important to remember that the endocrine system plays an important role [63]. The most commonly investigated biochemical markers in response to workload are testosterone and cortisol. Testosterone is an anabolic hormone that promotes amino acid incorporation into proteins whilst inhibiting protein breakdown [63]. Approximately 98% of testosterone is bound to carrier proteins such as sex-hormone-binding globulin (54%) and albumin and other proteins (44%) [64]. Of importance to practitioners is free testosterone, which is the part of serum testosterone that is available to tissues of the body [64]. Monitoring free testosterone levels can provide practitioners with an understanding of the anabolic status of the body [63]. Greater levels of free testosterone have been seen as a result of acute heavy resistance training [63]. However, conflicting findings have been reported in regards to the effect of training volume on resting free testosterone levels. A short-term investigation found a negative correlation between resting free testosterone levels with increases in training volume [65], whilst longitudinal studies have reported no changes in resting levels [66].

Cortisol is a catabolic hormone that converts amino acids to carbohydrates when muscle glycogen levels are depleted [63]. Similar to testosterone, there have been varied reports on the acute response of cortisol to workload with cortisol levels returning to pre-exercise levels within 2 to 3 h after cessation of exercise [67] whilst increased levels have also been observed for up to 24 h [67]. The free testosterone:cortisol (TC) ratio represents the imbalance between anabolic and catabolic state of the athlete or response to workload and has been used as a marker to determine anabolic and catabolic activity during periods of increased workloads [66,68,69]. During an 11-week training period in female weightlifters a very strong relationship was reported between percentage change in TC ratio and volume load ($r = -0.83$) and training intensity ($r = -0.72$) suggesting concomitant changes in the anabolic to catabolic ratio [17]. Specifically in regards to athlete monitoring, a decrease of 30% has been attributed towards overtraining with a number of investigations reporting significant relationships between performance and the TC ratio [8,66,69]. A longitudinal study involving basketball athletes over four consecutive years concluded that an athlete's hormonal status is linked with playing position, with power forwards and small forwards exhibiting the most catabolic state [70]. Overall, all players presented the most catabolic state in the final third of the regular season [70]. However, no changes were noted in testosterone, cortisol or TC ratio during a 28 day training camp in elite basketball athletes [71].

In addition to testosterone and cortisol, creatine kinase (CK) is also a commonly measured fatigue marker in athletes. The CK enzyme is stored inside muscle cells, however after heavy exercise is often released into the blood reflecting muscle damage [8]. Hence practitioners are interested in measuring CK levels to determine the level of exercise induced muscle damage. Acute CK responses have been documented in basketball [72] with increases following three days of tournament play. Longer investigations have also demonstrated increases in CK levels in team sport and non-team sport athletes. For example, following a six week deliberate overreaching phase in rugby league players, a significant increase in CK levels was observed [73]. Similar results were reported during six weeks of progressive endurance training in healthy adults [74]. The above evidence appears appealing for use of CK as a fatigue monitoring tool in basketball; however, large individual variability in resting CK levels exist which can make it problematic to measure change induced by training [75]. It is recommended that practitioners establish baseline levels for each athlete from a large number of samples in order to understand the degree of variability [8].

5. Fatigue Management

The advantages and disadvantages of the fatigue monitoring tools discussed in this review are outlined in Table 1 however practitioners can face unique challenges and scenarios depending on the time of year [8]. For example athletes are required to complete higher training volumes during a training camp or pre-season period. Strength and conditioning practitioners may use different strategies to manage athlete fatigue during this period compared to the competition period. The following sections discuss several methods in which the practitioner can manage athlete fatigue during intensive training camps and the competition period.

Table 1. Advantages and Disadvantages of Fatigue Monitoring Tools.

Fatigue Monitoring Tool	Advantages	Disadvantages
Vertical Jumps	• Easy to administer • Minimal additional fatigue • Replicates common athletic movement performed in competition • Easily implemented	• Lack of motivation to perform maximally • No consensus as to which variable is most sensitive to fatigue • Limited information regarding cause of performance reduction
Wellness Questionnaire	• No additional fatigue • Can be completed on a daily basis • Easy to administer	• Rely on subjective information • Athletes can manipulate data
Sprint Assessment	• Replicates movement performed in competition • Easily implemented • Provides information even when athlete not in a fatigue state	• May add to existing fatigue • Lack of motivation to perform maximally • Limited information regarding cause of performance reduction
Resting Heart Rate + Heart Rate Variability	• Most accessible physiological measure • Ability to capture over short period of time	• Valid for short term (<2 weeks) overload only • Limited evidence support use in team sports
Biochemical Markers	• Assist in understanding whether athlete is in a catabolic or anabolic state • CK levels may help determine level of muscle damage	• High time, cost and expertise demand for data collection • Time consuming analysis and feedback

5.1. Training Camp

Typically training camps or pre-season can last from seven weeks in collegiate basketball to only three weeks in the NBA. During this period athletes are exposed to high training loads to physically prepare them for the upcoming season. A challenge that faces strength and conditioning practitioners and coaching staff is the prescription of appropriate training volumes and recovery periods to optimize physiological adaptation and development of technical and tactical skills without the negative effects of high training loads [8]. Research indicates that players who complete a greater number of training sessions in the pre-season have a reduced injury rate during the competitive season [76]. Evidence also shows that teams with the lowest injury burdens had greater success in competition [77]. Whilst basketball training camps are typically shorter than those of other sports such as AF and rugby, it is recommended that strength and conditioning practitioners avoid large spikes (>10%) in workload to avoid increased risk of injury [5]. Given the shortened training camp in basketball compared to other sports, it is important to assess prior training load history as the off season break generally results in a low training base or chronic workload. A multi-disciplinary approach between sport coaches, strength and conditioning coaches, sport scientists, and athletes may be able to reduce the large spikes in workload period by voluntarily completing more training prior to the training camp, less training at the camp or a combination of both to ensure individual training prescription [5]. In addition, pairing individual athletes workload with fatigue monitoring tools will provide a global understanding of the dose-relationship and how the athlete is coping with the current workload [8]. Research studies report that no single fatigue monitoring tool can give a complete picture of an athlete's response to training and recommend using several fatigue monitoring tools across the squad of athletes to inform training and recovery decisions [14]. An example athlete monitoring system for training camp in basketball is detailed in Table 2.

Table 2. Example Monitoring System for a Basketball Training Camp [8].

Monitoring Tool	Frequency	Purpose	Analysis Method	Interpretation
Microtechnology (Player Load)	Every court based session	Measure of external load	• Z-score relative to individual • Acute to chronic ratio	• Z-score ≤ −1.5 • Acute to chronic ratio ≥1.5 = increased risk of injury
S-RPE training load	Every session	Measure of internal load	• Z-score relative to individual • Acute to chronic ratio	• Z-score ≤ −1.5 • Acute to chronic ratio ≥ 1.5 = increased risk of injury
Wellness Questionnaire	Daily	Measure of sleep quality, fatigue, soreness etc.	• Z-score to baseline measures • Smallest meaningful change relative to reliability	• Z-score ≤ −1.5 ± on item = positive or negative change
Countermovement Jump	Daily	Measure of neuromuscular fatigue	• Z-score to baseline measures • Smallest meaningful change relative to reliability	• Z-score ≤ −1.5 • If a variable decreases greater than the SWC
RHR/HRV	Daily	Measure of ANS	• Z-score to baseline measures	• Z-score ≤ −1.5

RHR = Resting Heart Rate; HRV = Heart Rate Variability; ANS = Autonomic Nervous System; S-RPE = Session Rating of Perceived Exertion; SWC = Smallest Worthwhile Change.

5.2. Competition Periods

A concern for practitioners during the competition period of basketball season is the impact of travel and the different turnaround times between matches. This must be considered by practitioners and sport coaches when planning the team's training program. Evidence in rugby league demonstrated that some positions had higher injury rates with longer turnaround times, whilst those in other positions had higher injury rates in shorter turnaround times [78]. Whilst there is limited research in basketball investigating the effects of travel and different turnaround times, practitioners need to take the positional differences into account as physical demands of training and competition are largely varied [23]. Quantifying individual athlete's workload and fatigue response can provide practitioners with insight in to how each athlete responds to travel, turnaround times and match load [8,78].

In any given week collegiate basketball teams play two games whilst in the NBA teams can play up to five games in a seven day period. Congested fixtures are also prevalent in post-season tournament play in which athletes are required to compete with only 24 h between games [8]. Evidence suggests higher levels of fatigue and increased injury rates are associated with congested fixtures due to spikes in game load [79]. Specifically, to basketball, 10 m sprint speed and CMJ height decreased until 24 (ES = 0.5) and 48 h (ES = 0.6) post-match [38]. These findings indicate that basketball athletes may need ~24–48 h of recovery post-match before the next intensive practice or match. Implementing fatigue monitoring strategies into daily training sessions such as CMJ and wellness questionnaires in combination with internal and external workload monitoring tools may assist practitioners to inform training and recovery strategies [8,38]. Longitudinally, athlete monitoring systems provide strength and conditioning practitioners a clearer understanding of scheduling variations and how each athlete responds to certain situations.

6. Interpretation Considerations

Before basketball practitioners use a fatigue monitoring tool it is important to ascertain the tool's reliability within their population as it has been shown to differ between sports and competition levels [8]. Several methods of assessing reliability of monitoring tools exist however intraclass correlations (ICC) and coefficient of variation (CV) are most common [8]. An ICC is used to determine the relationship between repeated tests or monitoring tools. A correlation of 1.0 represents a perfect relationship whilst 0.0 represents no relationship [8]. The CV refers to the typical error of a variable expressed as a percentage of the athlete's mean. A variable is often considered reliable when the ICC is >0.8 and/or if the CV is <10% [9]. In addition to establishing the reliability of a particular variable, the smallest worthwhile change (SWC) should also be calculated to allow practitioners to determine the smallest practical change in a fatigue monitoring tool that is important or worthwhile [8]. To calculate the SWC the following formula can be used: 0.2× between-subject standard deviation [8]. The SWC should be put into the context of the reliability of the fatigue monitoring tool. For example, for a practitioner to be confident that a change is not due to the noise associated with the test, the SWC should be greater than the CV [8]. However, it is important to incorporate both the CV and SWC together to determine the reliability and sensitivity of a fatigue monitoring tool variables that express the highest reliability may be too consistent and not sensitive to changes in athletic performance [8]. In contrast, variables that express poor reliability may be sensitive to fatigue despite having large variations that are greater than the SWC [8]. Therefore it is necessary that practitioners establishing the above statistics within their population in order to identify when athletes are in a fatigued state.

7. Conclusions

Basketball athletes playing at a professional or collegiate level participate in demanding pre-seasons to prepare for long playing seasons often coupled with extensive travel schedules. Ultimately this may result in an accumulation of perceptual and/or performance fatigue which could lead to a decrease in playing performance. Therefore, it is important that sport scientists and strength and conditioning practitioners implement appropriate athlete monitoring protocols to: (1) monitor the activity demands of training and competition; (2) monitor athlete fatigue levels; (3) prescribe appropriate recovery sessions; and (4) subsequently adjust and manage the athletes' workloads in order to potentially decrease and prevent high levels of fatigue that may affect playing performance. This review discussed several methods that may be used to quantify workload and athlete fatigue in basketball. However, it is important when practitioners pursue workload and fatigue monitoring tools that they also consider the feasibility, applicability and availability of equipment or resources.

Implementing an athlete monitoring program that includes workload and fatigue monitoring and management in basketball may assist practitioners and sport coaches to prescribe appropriate workloads that optimize training adaptations, decrease accumulated fatigue and allow athletes to perform at their highest level. Many of the discussed workload and fatigue monitoring tools have not been longitudinally investigated in basketball but are supported in several other sporting populations. Caution should be taken when initially implementing them into an athlete monitoring program. However, by implementing several workload and fatigue monitoring methods simultaneously, valuable information on an athlete's global fatigue and general workload trends over the length of a basketball season can be further understood. It should be noted that this review discusses common fatigue monitoring tools and variables reported in the literature that may be applied to monitor workload and fatigue in basketball. Several other methods can also be applied and incorporated in to an athlete monitoring program that may effectively monitor fatigue in basketball players that should also be investigated.

Author Contributions: T.E., B.P., T.S. and C.J. conceived and designed the review; T.E. performed the review and wrote the paper; B.P., T.S., J.B., G.G.H. and C.J. provided comments and edited the manuscript. All authors approved the final version of the article.

Conflicts of Interest: The authors declare no conflict of interest.

References

1. Stojanović, E.; Stojilijković, N.; Scanlan, A.T.; Dalbo, V.J.; Berkelmans, D.M.; Milanović, Z. The activity demands and physiological responses encountered during basketball match-play: A systematic review. *Sports Med.* **2017**. [CrossRef] [PubMed]
2. Abdelkrim, N.B.; El Fazaa, S.; El Ati, J. Time-motion analysis and physiological data of elite under-19-year-old basketball players during competition. *Br. J. Sports Med.* **2007**, *41*, 69–75. [CrossRef] [PubMed]
3. Scanlan, A.; Dascombe, B.; Reaburn, P. A comparison of the activity demands of elite and sub-elite Australian men's basketball competition. *J. Sport Sci.* **2011**, *29*, 1153–1160. [CrossRef] [PubMed]
4. Taylor, K.; Chapman, D.; Cronin, J.; Newton, M.J.; Gill, N. Fatigue monitoring in high performance sport: A survey of current trends. *J. Aust. Strength Cond.* **2012**, *20*, 12–23. [CrossRef]

5.   Drew, M.K.; Cook, J.; Finch, C.F. Sports-related workload and injury risk: Simply knowing the risks will not prevent injuries. *Br. J. Sports Med.* **2016**, *50*, 1306–1308. [CrossRef] [PubMed]
6.   Windt, J.; Gabbett, T.J. How do training and competition workloads relate to injury? The workload—injury aetiology model. *Br. J. Sports Med.* **2016**, *51*, 428–435. [CrossRef] [PubMed]
7.   Govus, A.D.; Coutts, A.; Duffield, R.; Murray, A.; Fullagar, H. Relationship between pre-training subjective wellness measures, player load and rating of perceived exertion training load in American college football. *Int. J. Sports Physiol. Perform.* **2017**, *10*, 1–19. [CrossRef]
8.   McGuigan, M. *Monitoring Training and Performance in Athletes*, 1st ed.; Human Kinetics: Champaign, IL, USA, 2017; pp. 1–201.
9.   Thorpe, R.T.; Atkinson, G.; Drust, B.; Gregson, W. Monitoring fatigue status in elite team sport athletes: Implications for practice. *Int. J. Sports Physiol. Perform.* **2017**, *12*, 27–34. [CrossRef] [PubMed]
10. Thorpe, R.T.; Strudwick, A.J.; Buchheit, M.; Atkinson, G.; Drust, B.; Gregson, W. Tracking morning fatigue status across in-season training weeks in elite soccer players. *Int. J. Sports Physiol Perform.* **2016**, *11*, 947–952. [CrossRef] [PubMed]
11. Enoka, R.M.; Duchateau, J. Translating fatigue to human performance. *Med. Sci. Sports Exerc.* **2016**, *48*, 2228–2238. [CrossRef] [PubMed]
12. Coutts, A.J.; Reaburn, P.; Piva, T.J.; Rowsell, G.J. Monitoring for overreaching in rugby league players. *Eur. J. Appl. Physiol.* **2007**, *99*, 313–324. [CrossRef] [PubMed]
13. Kraemer, W.J.; French, D.N.; Paxton, N.J.; Volek, J.S.; Sebastianelli, W.J.; Putukian, M.; Newton, R.U.; Rubin, M.R.; Gomez, A.J.; Vascovi, J.D. Changes in exercise performance and hormonal concentrations over a big ten soccer season in starters and nonstarters. *J. Strength Cond. Res.* **2004**, *18*, 121–128. [CrossRef] [PubMed]
14. Buchheit, M.; Racinais, S.; Bilsborough, J.C.; Bourdon, P.C.; Voss, S.C.; Hocking, J.; Cordy, J.; Mendez-Villanueva, A.; Coutts, A.J. Monitoring fitness, fatigue and running performance during a pre-season training camp in elite football players. *J. Sci. Med. Sport.* **2013**, *16*, 550–555. [CrossRef] [PubMed]
15. Gallo, T.F.; Cormack, S.J.; Gabbett, T.J.; Lorenzen, C.H. Pre-training perceived wellness impacts training output in Australian football players. *J. Sports Sci.* **2016**, *34*, 1445–1451. [CrossRef] [PubMed]
16. Gastin, P.B.; Meyer, D.; Robinson, D. Perceptions of wellness to monitor adaptive responses to training and competition in elite Australian football. *J. Strength Cond. Res.* **2013**, *27*, 2518–2526. [CrossRef] [PubMed]
17. Haff, G.G.; Jackson, J.R.; Kawamori, N.; Carlock, J.M.; Hartman, M.J.; Kilgore, J.L.; Morris, R.T.; Ramsey, M.W.; Sands, W.A.; Stone, M.H. Force-time curve characteristics and hormonal alterations during an eleven-week training period in elite women weightlifters. *J. Strength Cond. Res.* **2008**, *22*, 433–446. [CrossRef] [PubMed]
18. Cormack, S.J.; Newton, R.U.; McGuigan, M.R.; Cormie, P. Neuromuscular and endocrine responses of elite players during an Australian rules football season. *Int. J. Sports Physiol. Perform.* **2008**, *3*, 439–453. [CrossRef] [PubMed]
19. Manzi, V.; D'ottavio, S.; Impellizzeri, F.M.; Chaouachi, A.; Chamari, K.; Castagna, C. Profile of weekly training load in elite male professional basketball players. *J. Strength Cond. Res.* **2010**, *24*, 1399–1406. [CrossRef] [PubMed]
20. Fox, J.L.; Scanlan, A.T.; Stanton, R. A review of player monitoring approaches in basketball: Current trends and future directions. *J. Strength Cond. Res.* **2017**, *31*, 2021–2029. [CrossRef] [PubMed]
21. Cummins, C.; Orr, R.; O'Connor, H.; West, C. Global positioning systems (GPS) and microtechnology sensors in team sports: a systematic review. *Sports Med.* **2013**, *43*, 1025–1042. [CrossRef] [PubMed]
22. Malone, J.J.; Lovell, R.; Varley, M.C.; Coutts, A.J. Unpacking the black box: Applications and considerations for using GPS devices in sport. *Int. J. Sports Physiol. Perf.* **2016**, *12*, 1–30. [CrossRef] [PubMed]
23. Schelling, X.; Torres-Ronda, L. Accelerometer load profiles for basketball-specific drills in elite players. *J. Sports Sci. Med.* **2016**, *15*, 585–591. [PubMed]
24. Fox, J.L.; Stanton, R.; Scanlan, A.T. A comparison of training and competition demands in semiprofessional male basketball players. *Res. Q. Exerc. Sport.* **2018**, *89*, 103–111. [CrossRef] [PubMed]
25. Foster, C. Monitoring training in athletes with reference to overtraining syndrome. *Med. Sci. Sports Exerc.* **1998**, *30*, 1164–1168. [CrossRef]
26. Foster, C.; Florhaug, J.A.; Franklin, J.; Gottschall, L.; Hrovatin, L.A.; Parker, S.; Doleshal, P.; Dodge, C. A new approach to monitoring exercise training. *J. Strength Cond. Res.* **2001**, *15*, 109–115. [CrossRef] [PubMed]

27. Alexiou, H.; Coutts, A.J. A comparison of methods used for quantifying internal training load in women soccer players. *Int. J. Sports Physiol. Perf.* **2008**, *3*, 320–330. [CrossRef]
28. Coutts, A.J.; Reaburn, P. Monitoring changes in rugby league players' perceived stress and recovery during intensified training. *Percept. Mot. Skills* **2008**, *106*, 904–916. [CrossRef] [PubMed]
29. Klusemann, M.J.; Pyne, D.B.; Hopkins, W.G.; Drinkwater, E.J. Activity profiles and demands of seasonal and tournament basketball competition. *Int. J. Sports Physiol. Perf.* **2013**, *8*, 623–629. [CrossRef]
30. Moreira, A.; McGuigan, M.R.; Arruda, A.F.; Freitas, C.G.; Aoki, M.S. Monitoring internal load parameters during simulated and official basketball matches. *J. Strength Cond. Res.* **2012**, *26*, 861–866. [CrossRef] [PubMed]
31. Montgomery, P.G.; Pyne, D.B.; Minahan, C.L. The physical and physiological demands of basketball training and competition. *Int. J. Sports Physiol. Perf.* **2010**, *5*, 75–86. [CrossRef]
32. Scanlan, A.T.; Wen, N.; Tucker, P.S.; Dalbo, V.J. The relationships between internal and external training load models during basketball training. *J. Strength Cond. Res.* **2014**, *28*, 2397–2405. [CrossRef] [PubMed]
33. Dellal, A.; Keller, D.; Carling, C.; Chaouachi, A.; Wong del, P.; Chamari, K. Physiologic effects of directional changes in intermittent exercise in soccer players. *J. Sterngth Cond Res.* **2010**, *24*, 3219–3226. [CrossRef] [PubMed]
34. McInnes, S.; Carlson, J.S.; Jones, C.J.; McKenna, M.J. The physiological load imposed on basketball players during competition. *J. Sports Sci.* **1995**, *13*, 387–397. [CrossRef] [PubMed]
35. Dawes, J.J.; Spiteri, T. Relationship between pre-season testing performance and playing time among NCAA DII basketball players. *Sports Exerc. Med.* **2016**, *2*, 47–54. [CrossRef]
36. Hoffman, J.R.; Tenenbaum, G.; Maresh, C.M.; Kraemer, W.J. Relationship between athletic performance tests and playing time in elite college basketball players. *J. Strength Cond. Res.* **1996**, *10*, 67–71. [CrossRef]
37. McGill, S.M.; Andersen, J.T.; Horne, A.D. Predicting performance and injury resilience from movement quality and fitness scores in a basketball team over 2 years. *J. Strength Cond. Res.* **2012**, *26*, 1731–1739. [CrossRef] [PubMed]
38. Chatzinikolaou, A.; Draganidis, D.; Avloniti, A.; Karipidis, A.; Jamurtas, A.Z.; Skevaki, C.L.; Tsoukas, D.; Sovatzidis, A.; Theodorou, A.; Kambas, A.; et al. The microcycle of inflammation and performance changes after a basketball match. *J. Sports Sci.* **2014**, *32*, 870–882. [CrossRef] [PubMed]
39. Buchheit, M. Sensitivity of monthly heart rate and psychometric measures for monitoring physical performance in highly trained young handball players. *Int. J. Sports Med.* **2015**, *36*, 351–356. [CrossRef] [PubMed]
40. Coutts, A.J.; Slattery, K.M.; Wallace, L.K. Practical tests for monitoring performance, fatigue and recovery in triathletes. *J. Sci. Med. Sport.* **2007**, *10*, 372–381. [CrossRef] [PubMed]
41. Kenttä, G.; Hassmén, P. Overtraining and recovery. *Sports Med.* **1998**, *26*, 1–16. [CrossRef] [PubMed]
42. Gallo, T.F.; Cormack, S.J.; Gabbett, T.J.; Lorenzen, C.H. Self-reported wellness profiles of professional Australian football players during the competition phase of the season. *J. Strength Cond. Res.* **2017**, *31*, 495–502. [CrossRef] [PubMed]
43. Komi, P.V. Stretch-shortening cycle: A powerful model to study normal and fatigued muscle. *J. Biomech.* **2000**, *33*, 1197–1206. [CrossRef]
44. Nicol, C.; Avela, J.; Komi, P.V. The stretch-shortening cycle. *Sports Med.* **2006**, *36*, 977–999. [CrossRef] [PubMed]
45. Legg, J.; Pyne, D.B.; Semple, S.; Ball, N. Variability of jump kinetics related to training load in elite female basketball. *Sports* **2017**, *5*, 85. [CrossRef]
46. Markwick, W. Training Load Quantification in Professional Australian Basketball and the Use of the Reactive Strength Index as a Monitoring Tool. Master's Thesis, Edith Cowan University, Perth, Australia, 2015.
47. Claudino, J.G.; Cronin, J.; Mezencio, B.; McMaster, D.T.; McGuigan, M.; Tricoli, V.; Amadio, A.C.; Serrao, J.C. The countermovement jump to monitor neuromuscular status: A meta-analysis. *J. Sci. Med. Sport.* **2017**, *20*, 397–402. [CrossRef] [PubMed]
48. Ronglan, L.; Raastad, T.; Børgesen, A. Neuromuscular fatigue and recovery in elite female handball players. *Scand. J. Med. Sci. Sports.* **2006**, *16*, 267–273. [CrossRef] [PubMed]
49. Gibson, N.E.; Boyd, A.J.; Murray, A.M. Countermovement jump is not affected during final competition preparation periods in elite rugby sevens players. *J. Strength Cond. Res.* **2016**, *30*, 777–783. [CrossRef] [PubMed]

50. Gathercole, R.; Sporer, B.; Stellingwerff, T.; Sleivert, G. Alternative countermovement-jump analysis to quantify acute neuromuscular fatigue. *Int. J. Sports Physiol. Perform.* **2015**, *10*, 84–92. [CrossRef] [PubMed]
51. Spiteri, T.; Nimphius, S.; Wolski, A.; Bird, S. Monitoring neuromuscular fatigue in female basketball players across training and game performance. *J. Aust. Strength Cond.* **2013**, *21*, 73–74.
52. Spiteri, T.; Nimphius, S.; Hart, N.H.; Specos, C.; Sheppard, J.M.; Newton, R.U. Contribution of strength characteristics to change of direction and agility performance in female basketball athletes. *J. Strength Cond. Res.* **2014**, *28*, 2415–2423. [CrossRef] [PubMed]
53. Achten, J.; Jeukendrup, A.E. Heart rate monitoring. *Sports Med.* **2003**, *33*, 517–538. [CrossRef] [PubMed]
54. Bosquet, L.; Merkari, S.; Arvisais, D.; Aubert, A.E. Is heart rate a convenient tool to monitor over-reaching? A systematic review of the literature. *Br. J. Sports Med.* **2008**, *42*, 709–714. [CrossRef] [PubMed]
55. Dressendorfer, R.H.; Wade, C.E.; Scaff, J.H., Jr. Increased morning heart rate in runners: A valid sign of overtraining? *Phys. Sportsmed.* **1985**, *13*, 77–86. [CrossRef] [PubMed]
56. Fry, A.C.; Kraemer, W.J.; van Borselen, F.; Lynch, J.M.; Marsit, J.L.; Roy, E.P.; Triplett, N.T.; Knuttgen, H.G. Performance decrements with high-intensity. *Med. Sci. Sports Exerc.* **1994**, *26*, 1165–1173. [CrossRef]
57. Fry, R.W.; Morton, A.R.; Keast, D. Overtraining in athletes. *Sports Med.* **1991**, *12*, 32–65. [CrossRef] [PubMed]
58. Al Haddad, H.; Laursen, P.B.; Chollet, D.; Ahmaidi, S.; Bucheit, M. Reliability of resting and postexercise heart rate measures. *Int. J. Sports Med.* **2011**, *32*, 598–605. [CrossRef] [PubMed]
59. Esco, M.R.; Flatt, A.A. Ultra-short-term heart rate variability indexes at rest and post-exercise in athletes: evaluating the agreement with accepted recommendations. *J. Sports Sci. Med.* **2014**, *13*, 535–541. [PubMed]
60. Chen, J.L.; Yeh, D.P.; Lee, J.P.; Huang, C.Y.; Lee, S.D.; Chen, C.C.; Kuo, T.B.; Kao, C.L.; Kuo, C.H. Parasympathetic nervous activity mirrors recovery status in weightlifting performance after training. *J. Strength Cond. Res.* **2011**, *25*, 1546–1552. [CrossRef] [PubMed]
61. Atlaoui, D.; Pichot, V.; Lacoste, L.; Barale, F.; Lacour, J.R.; Chatard, J.C. Heart rate variability, training variation and performance in elite swimmers. *Int. J. Sports Med.* **2007**, *28*, 394–400. [CrossRef] [PubMed]
62. Pichot, V.; Roche, F.; Gaspoz, J.M.; Barthelemy, J.C. Relation between heart rate variability and training load in middle-distance runners. *Med. Sci. Sports Exerc.* **2000**, *32*, 1729–1736. [CrossRef] [PubMed]
63. Haff, G.G.; Triplett, N.T. *Essentials of Strength Training and Conditioning*, 4th ed.; Human Kinetics: Champaigne, IL, USA, 2015; ISBN 978-1-4925-0162-6.
64. Hug, M.; Mullis, P.E.; Vogt, M.; Ventura, N.; Hoppeler, H. Training modalities: Over-reaching and over-training in athletes, including a study of the role of hormones. *Best Pract. Res. Clin. Endocrinol. Metab.* **2003**, *17*, 191–209. [CrossRef]
65. Häkkinen, K.; Pakarinen, A.; Alen, M.; Kauhanen, H.; Komi, P.V. Daily hormonal and neuromuscular responses to intensive strength training in 1 week. *Int. J. Sports Med.* **1988**, *9*, 422–428. [CrossRef] [PubMed]
66. Häkkinen, K.; Pakarinen, A.; Alen, M.; Kauhanen, H.; Komi, P.V. Relationships between training volume, physical performance capacity, and serum hormone concentrations during prolonged training in elite weight lifters. *Int. J. Sports Med.* **1987**, *8*, S61–S65. [CrossRef]
67. Maron, M.B.; Horvath, S.M.; Wilkerson, J.E. Blood biochemical alterations during recovery from competitive marathon running. *Eur. J. Appl. Physiol.* **1977**, *36*, 231–238. [CrossRef]
68. Cormack, S.J.; Newton, R.U.; McGuigan, M. Neuromuscular and endocrine responses of elite players to an Australian rules football match. *Int. J. Sports Physiol. Perform.* **2008**, *3*, 359–374. [CrossRef] [PubMed]
69. Banfi, G.; Marinelli, M.; Roi, G.S.; Agape, V. Usefulness of free testosterone/cortisol ratio during a season of elite speed skating athletes. *Int. J. Sports Med.* **1993**, *14*, 373–379. [CrossRef] [PubMed]
70. Schelling, X.; Calleja-Gonzalez, J.; Torres-Rinda, L.; Terrados, N. Using testosterone and cortisol as biomarker for training individualization in elite basketball: A 4-year follow-up study. *J. Strength Cond. Res.* **2015**, *29*, 368–378. [CrossRef] [PubMed]
71. Hoffman, J.R.; Epstein, S.; Yarom, Y.; Einbinder, M. Hormonal and biochemical changes in elite basketball players during a 4-week training camp. *J. Strength Cond. Res.* **1999**, *13*, 280–285. [CrossRef]
72. Montgomery, P.G.; Pyne, D.B.; Cox, A.J.; Hopkins, W.G.; Minahan, C.L.; Hunt, P.H. Muscle damage, inflammation, and recovery interventions during a 3-day basketball tournament. *Eur. J. Sports Sci.* **2008**, *8*, 241–250. [CrossRef]
73. Coutts, A.; Reaburn, P.; Piva, T.J.; Murphy, A. Changes in selected biochemical, muscular strength, power, and endurance measures during deliberate overreaching and tapering in rugby league players. *Int. J. Sports Med.* **2007**, *28*, 116–124. [CrossRef] [PubMed]

74. Kargotich, S.; Keast, D.; Goodman, C.; Morton, A.R. Monitoring 6 weeks of progressive endurance training with plasma glutamine. *Int. J. Sports Med.* **2007**, *28*, 211–216. [CrossRef] [PubMed]

75. Hartmann, U.; Mester, J. Training and overtraining markers in selected sport events. *Med. Sci. Sports Exerc.* **2000**, *32*, 209–215. [CrossRef] [PubMed]

76. Windt, J.; Gabbett, T.J.; Ferris, D.; Khan, K.M. Training load—Injury paradox: Is greater preseason participation associated with lower in-season injury risk in elite rugby league players? *Br. J. Sports Med.* **2016**, *51*, 645. [CrossRef] [PubMed]

77. Hägglund, M.; Walden, M.; Magnusson, H.; Kristenson, K.; Bengtsson, H.; Ekstand, J. Injuries affect team performance negatively in professional football: An 11-year follow-up of the UEFA Champions League injury study. *Br. J. Sports Med.* **2013**, *47*, 738–742. [CrossRef] [PubMed]

78. Murray, N.B.; Gabbett, T.J.; Chamari, K. Effect of different between-match recovery times on the activity profiles and injury rates of national rugby league players. *J. Strength Cond. Res.* **2014**, *28*, 3476–3483. [CrossRef] [PubMed]

79. Dellal, A.; Lago-Penas, C.; Rey, E.; Chamari, K.; Orhant, E. The effects of a congested fixture period on physical performance, technical activity and injury rate during matches in a professional soccer team. *Br. J. Sports Med.* **2013**, *49*, 390. [CrossRef] [PubMed]

© 2018 by the authors. Licensee MDPI, Basel, Switzerland. This article is an open access article distributed under the terms and conditions of the Creative Commons Attribution (CC BY) license (http://creativecommons.org/licenses/by/4.0/).

sports

MDPI

Article

Team Performance Indicators Explain Outcome during Women's Basketball Matches at the Olympic Games

Anthony S. Leicht [1,*] (ID), Miguel A. Gomez [2] and Carl T. Woods [1]

[1]   Sport and Exercise Science, James Cook University, Townsville 4811, Australia; Carl.Woods@jcu.edu.au
[2]   Faculty of Physical Activity and Sport Sciences, Polytechnic University of Madrid, Madrid, Spain;
      magor_2@yahoo.es
*   Correspondence: Anthony.Leicht@jcu.edu.au; Tel.: +61-7-47814576

Received: 19 October 2017; Accepted: 7 December 2017; Published: 17 December 2017

Abstract: The Olympic Games is the pinnacle international sporting competition with team sport coaches interested in key performance indicators to assist the development of match strategies for success. This study examined the relationship between team performance indicators and match outcome during the women's basketball tournament at the Olympic Games. Team performance indicators were collated from all women's basketball matches during the 2004–2016 Olympic Games ($n = 156$) and analyzed via linear (binary logistic regression) and non-linear (conditional interference (CI) classification tree) statistical techniques. The most parsimonious linear model retained "defensive rebounds", "field-goal percentage", "offensive rebounds", "fouls", "steals", and "turnovers" with a classification accuracy of 85.6%. The CI classification tree retained four performance indicators with a classification accuracy of 86.2%. The combination of "field-goal percentage", "defensive rebounds", "steals", and "turnovers" provided the greatest probability of winning (91.1%), while a combination of "field-goal percentage", "steals", and "turnovers" provided the greatest probability of losing (96.7%). Shooting proficiency and defensive actions were identified as key team performance indicators for Olympic female basketball success. The development of key defensive strategies and/or the selection of athletes highly proficient in defensive actions may strengthen Olympic match success. Incorporation of non-linear analyses may provide teams with superior/practical approaches for elite sporting success.

Keywords: team sports; classification tree; machine learning; performance analysis; non-linear analysis; athlete

1. Introduction

Basketball is the second most popular team sport worldwide, and the second most watched Olympic sport, with over 450 million registered players [1]. The key physical and physiological characteristics of basketball athletes have been documented [2–4] and reported to contribute to individual performance [5,6] with team success reliant on the coherent integration of individual performances [7,8]. Many studies have examined the importance of team performance indicators for match success within national junior and senior competitions [9–13]. Most have identified "field-goal percentage", "defensive rebounds", and "assists" as crucial team indicators for match success [9,11–14]. Recently, these results were extended to the elite international level with "field goal percentage" and "defensive rebounds" identified as vital for match outcomes at the men's Olympic basketball tournaments [8]. Key team performance was identified using both linear and non-linear statistical techniques with a classification tree (a non-linear machine learning technique) providing coaches with a practical guide to inform match strategy and team selection [8]. Recently, others have utilized a classification and regression tree to identify predictors of winning within the Spanish EBA Basketball League [15]. Similar analytical approaches have been applied in elite Australian Football [16] and rugby league [17] and highlight these novel techniques as important tools for sport scientists and coaches to improve on-field success. However, most studies examining team performance indicators and match success have focused on male athletes [9–14,16,17] with very few studies examining females [18–20], and less for basketball [18–20]. Performance differences due to dissimilarities in anthropometrical and fitness characteristics between male and female basketball athletes [21] may impact on game-related statistics and match success [18]. Subsequently, it is critical to examine predictors of match success for elite female basketball athletes for a greater understanding of factors contributing to match success.

To date, only two studies have focused on female basketball athletes and match performance indicators for success. Gomez et al. [19] examined matches within the 2004/2005 women's Spanish professional league and demonstrated that "1-point" and "3-point field-goal percentage", "assists", and "defensive rebounds" were important during balanced games (score-differences ≤12 points) and "2-point field-goal percentages", "defensive rebounds", and "steals" during unbalanced games (score-differences >12 points). In a second study, Gomez et al. [7] examined the impact of starter/nonstarter player status, team performance indicators, and match outcome within the 2005 Women's National Basketball Association. These authors reported that shooting (2-point field-goals, successful free-throws) and passing capability (assists) were discriminatory of player status with this profile impacting on match success. Recently, others examined match performance in elite, female Spanish basketball with steals and assists correlated with a range of physical fitness characteristics (e.g., speed, agility, anaerobic power, repeated sprint ability, and aerobic power) [6]. To our knowledge, no other studies have examined match outcome and team performance indicators for female basketball players. Identification of the relationship between team performance indicators and match success, particularly at the elite level, would provide significant guidance to coaches and athletes in the development of training and match strategies for match success.

The aim of the current study was to identify the relationship between team performance indicators and match outcome during elite women's basketball competition using linear and non-linear statistical techniques. Based on previous results [8], it was hypothesized that distinctive performance indicator combinations would explain match outcome with the non-linear technique, offering greater practical utility for coaches and athletes.

2. Materials and Methods

This study was a retrospective analysis of publically available data from the official Olympic websites. All matches (n = 156) undertaken within the women's basketball tournament at the past four Olympic tournaments (2016, Rio de Janeiro, n = 38; 2012, London, n = 38; 2008, Beijing, n = 38; 2004, Athens, n = 42) were examined. As previously described [8], team performance indicators ('field-goal percentage", "3-point percentage", "2-point percentage", "free-throw percentage", "offensive rebounds", "defensive rebounds", "assists", "turnovers", "steals", "blocked shots", "fouls committed", and "fouls against") were downloaded, collated, and *a priori* classified according to match outcome (win/loss). Normalization of all team performance indicators was undertaken using the number of ball possessions, as previously described [8,13,22]. Two datasets (one per team) were obtained from each match with 312 datasets (76 from 2016, 76 from 2012, 76 from 2008, 84 from 2004) examined in the current study.

Relative to match outcome, descriptive statistics (mean ± SD) were calculated for each team performance indicator with all analyses and visualizations conducted using R (version 3.2.2, Vienna, Austria). Match outcome comparisons of each performance indicator were examined via multivariate analysis of variance (MANOVA) with the level of statistical significance set at $p < 0.05$. The magnitude of effect (i.e., effect size and 90% confidence intervals) for match outcome comparisons were calculated using Cohen's d statistic as follows: $d < 0.2$: trivial; $d = 0.20$–0.49: small; $d = 0.50$–0.79: medium; $d > 0.79$: large [23].

As described previously [8], data were examined via both binary logistic regression and a conditional interference (CI) classification tree. Briefly, match outcome was coded as the response variable with each identified performance indicator coded as the explanatory variable within both statistical techniques. Model parsimony for the binary logistic regression was performed using the delta Akaike Information Criterion (AIC) and Akaike weights [24] via the "dredge" function in the MuMIn package [24]. A null model was built and used as a comparator. A recursively, partitioned CI, classification tree was grown via the "ctree" function in the party package [25] with a minimum node size of 5 observations chosen for partitioning. This type of classification tree was chosen as its fitting algorithm corrects for multiple testing, thus avoiding overfitting [25]. Accordingly, this analysis results in the growth of an unbiased decision tree that does not require pruning [25].

3. Results

During wins, all of the normalized, team performance indicators were significantly greater than for losses, with the exception for "turnovers", which was significantly lower, and "fouls committed", which was similar (Table 1). The indicators that had the largest effect on match outcome were "field-goal percentage", "defensive rebounds", "assists", and "steals" (Table 1).

Table 1. Descriptive statistics for each team performance indicator relative to match outcome. Values are mean $\pm$ SD with each normalized to ball possessions.

Performance Indicator	Wins	Losses	d (90% CI)	Interpretation
Field-goal percentage	77.9 $\pm$ 13.8	60.6 $\pm$ 12.8 *	1.30 (1.09, 1.50)	Large
Free-throw percentage	129.4 $\pm$ 22.1	117.6 $\pm$ 23.0 *	0.52 (0.33, 0.71)	Medium
Offensive rebounds	22.2 $\pm$ 8.4	17.4 $\pm$ 8.7 *	0.55 (0.36, 0.74)	Medium
Defensive rebounds	47.4 $\pm$ 9.7	35.9 $\pm$ 9.2 *	1.21 (1.00, 1.41)	Large
Assists	27.9 $\pm$ 10.4	19.2 $\pm$ 8.5 *	0.91 (0.71, 1.10)	Large
Turnovers	25.7 $\pm$ 8.0	28.4 $\pm$ 7.5 *	-0.35 (-0.54, -0.16)	Small
Steals	15.5 $\pm$ 5.4	10.8 $\pm$ 5.1 *	0.90 (0.71, 1.10)	Large
Blocked shots	5.7 $\pm$ 3.8	3.4 $\pm$ 2.9 *	0.66 (0.47, 0.85)	Medium
Fouls committed	30.4 $\pm$ 8.5	31.4 $\pm$ 8.2	-0.13 (-0.32, 0.06)	Small
Fouls against	33.2 $\pm$ 10.1	29.3 $\pm$ 9.0 *	0.41 (0.23, 0.60)	Small

$n = 312$; * $p < 0.005$ vs. Wins; d—effect size; CI—confidence interval.

The following performance indicators were retained by the best linear model: "defensive rebounds", "field-goal percentage", "offensive rebounds", "fouls", "steals", and "turnovers" (Table 2). This model successfully identified 88.5% and 89.1% of the *a priori* classified wins and losses, respectively, for an average model accuracy of 85.6%.

Four performance indicators were retained within the CI classification tree (Figure 1) with the tree successfully classifying 94.2% and 78.2% of the *a priori* classified wins and losses, respectively, for an average model accuracy of 86.2%. The root node (Number 1) partitioned the dataset based on "field-goal percentage" and generated eight terminal nodes (Numbers 4, 5, 6, 8, 11, 12, 14, and 15). The left-hand branching of the tree denoted primarily a loss (field-goal percentage $\leq$62.243), while the branching to the right primarily denoted a win (field-goal percentage >62.243).

On the left-hand side of the tree, Node Number 2 separated the data based on "steals" and generated Terminal Node 6, while Node Number 3 further separated the data based on "turnovers" to generate Terminal Nodes 4 and 5. The combination of "field-goal percentage" ($\leq$62.243%), "steals" ($\leq$18.841), and "turnovers" (>18.362) provided the greatest probability of losing (96.7%, Terminal Node 5).

On the right-hand side of the tree, Node Number 7 separated the data based on "defensive rebounds" and generated Terminal Node 8, while Node Number 9 further separated the data based on "steals" to generate Nodes 10 and 13. Finally, Nodes 10 and 13 split the data based upon by "defensive rebounds" and "turnovers," respectively. The combination of "field-goal percentage" (>62.243%), "defensive rebounds" (>30.789), "steals" (>9.317), and "turnovers" (<36.63) provided the greatest probability of winning (91.1%, Terminal Node 14).

Table 2. Model summary for the binary logistic regression analysis ranked according to the delta Akaike Information Criterion and Akaike weights.

Predictors	LL	df	AICc	ΔAIC	w_i
~def_reb + field_goal + off_reb + fouls + steals + turnovers	−82.93	7	180.23	<0.01	0.15
~blocked_shots + def_reb + field_goal + fouls + off_reb + steals + turnovers	−82.50	8	181.47	1.24	0.08
~def_reb + field_goal + fouls + steals + turnovers	−84.68	6	181.63	1.40	0.07
~def_reb + field_goal + fouls + free_throw + off_reb + steals + turnovers	−82.72	8	181.93	1.70	0.06
~assists + def_reb + field_goal + fouls + off_reb + steals + turnovers	−82.88	8	182.24	2.01	0.05
~def_reb + field_goal + fouls + fouls_against + off_reb + steals + turnovers	−82.92	8	182.31	2.08	0.05
~blocked_shots + def_reb + field_goal + fouls + steals + turnovers	−84.03	7	182.43	2.20	0.05
~blocked_shots + def_reb + field_goal + fouls + free_throw + off_reb + steals + turnovers	−82.27	9	183.13	2.90	0.04
Null (~1)	−216.26	1	434.54	254.31	<0.01

LL: log likelihood; *df*: degrees of freedom; AICc: Akaike Information Criterion; ΔAIC: delta AIC; w_i: Akaike weight; def_reb: defensive rebounds; field_goal: field goal percentage; off_reb: offensive rebounds; free_throw: free-throw percentage.

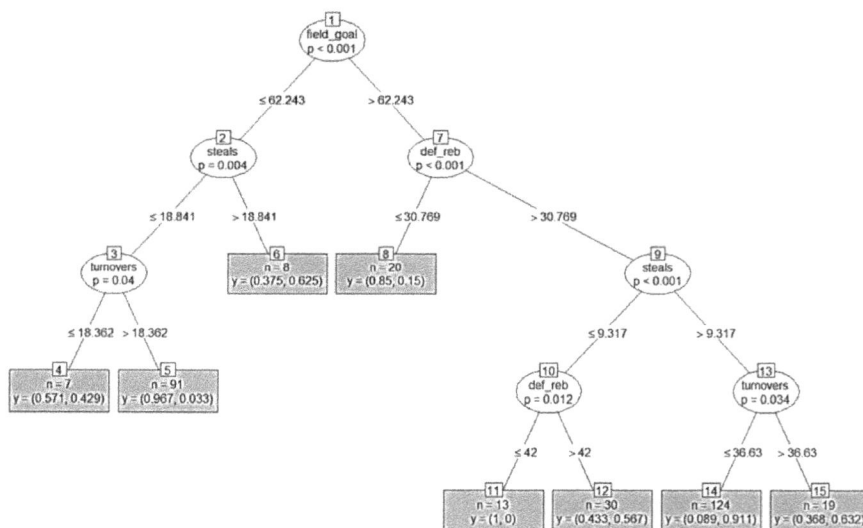

Figure 1. The conditional interference classification tree highlighting the probability of wins and losses during the women's basketball tournament of the 2004–2016 Olympic Games. "*n*" denotes the number of observations or datasets in each node (minimum of 5) with the first y-value denoting the probability of losing and the second y-value denoting the probability of winning (e.g., 0.7 = 70%). field_goal = "field-goal percentage"; def_reb = "defensive rebounds"; values for each team performance indicators were normalized to ball possessions.

4. Discussion

The current study identified the key team performance indicators that contributed to success in women's basketball at the 2004–2016 Olympic Games. The non–linear analysis resolved a combination of "field-goal percentage", "defensive rebounds", "steals", and "turnovers" as providing the greatest probability of winning (91.1%). Further, a unique combination of "field-goal percentage", "steals", and "turnovers" offered the lowest probability of winning (3.3%) and the greatest probability of losing

(96.7%). Overall, the average model accuracy was marginally higher for the CI classification tree compared with the logistic regression analysis and likely provided coaches and analysts with a flexible model to manipulate game plans or strategies to enhance the likelihood of winning. The use of non-linear, machine learning techniques may provide sport scientists with greater support when assisting coaches with decisions regarding match strategy design, team selection or identifying opponent strengths and weaknesses [8,15].

In our previous work, "field-goal percentage", "defensive rebounds", "steals", and "turnovers" were identified as key indicators of outcome for men's matches at the Olympic Games [8]. The current results extend these findings to women's matches at the Olympic Games and confirm these indicators as significant, sex-independent contributors to basketball success at the current Olympic level. Further, our results for Olympic matches highlight shooting proficiency and defensive actions as vital for success in both men's and women's basketball. Others have reported the importance of "field-goal percentage" [7,9,13,14,20,26], "defensive rebounds" [9,10,12–15,20,26] and "turnovers" [13,14] for basketball match success in various competitions. While shooting capability may seem apparent for match success, particularly longer distance shots for females [19], collectively the current and prior results [9,10,12–14,20,26] confirmed defensive actions as critical for match success. Gomez et al. [9] identified "defensive rebounds" as the predominant performance indicator to discriminate winning and losing within the Spanish Men's Basketball League. Similarly, Trninic et al. [14] identified "defensive rebounds" as the key discriminator for success at the European club championships. These authors commented that winning teams exhibited a greater discipline and balance of play highlighted by greater decision making and teamwork [14]. Subsequently, inclusion of athletes that are familiar with each other and a controlled style of play [15], or who are more tactically disciplined [22], may provide greater defensive actions for match success. This degree of familiarity, focus, and discipline may be difficult given the limited preparation time and match opportunities for national teams that include athletes competing potentially in every corner of the world. Therefore, preparatory activities for individual athletes, development of team cohesion and a focus on team defensive activities may be vital for Olympic success. Increasing defensive pressure on the opposition was reported to reduce basketball athlete's preference to shoot [27] that may provide further impact on shooting proficiency and overall match success. Coaches are encouraged to develop key defensive strategies and/or selection of athletes highly proficient in defensive actions for greater Olympic match success.

A key finding of the current study was the substantial effect of "steals" on match success. This result extended our previous finding that "steals" was a key performance indicator for elite Olympic basketball success [8] and unbalanced games within the Spanish Women's League [19], and confirms "steals" as an important focus area for coaches and athletes. Interestingly, "steals" were reported to differentiate men's and women's teams within an analysis of close matches during the basketball World Championships in 1999–2002 [18]. Compared to women's teams, men's teams were associated with a lower proportion of "steals" which was related to their anthropometric characteristics (i.e., taller and heavier) [18]. Our current and previous [8] results indicate similar sex differences for "steals" during wins (women's = 15.5 vs. men's = 10.5) but highlight further the importance of this indicator for match success, independent of sex. Subsequently, coaches of men's and women's teams are encouraged to develop strategies to enhance "steals" during elite basketball matches. These strategies may include full- and half-court presses and double teaming of players to enhance the likelihood of match success [22,28]. Further, selection of athletes that possess superior fitness characteristics may be important for the generation of "steals." Previously, "steals" were associated with superior speed, agility, anaerobic power, and repeated sprint ability in elite, junior, female basketball athletes [6]. The inclusion of athletes that possess these characteristics for national teams may provide the talent base to enhance "steals" during matches and ultimate success. Additionally, development of these fitness characteristics, specific to each position [5], during the pre-Olympic period may be suggested as a priority for coaches in their preparation for the games [29].

The current study has expanded the understanding of match success for Olympic women's basketball competition. Through the use of linear and non-linear statistical techniques, key performance indicators were identified to assist coaches and athletes in their preparation for international, women's basketball competition. However, some limitations of the current study should be discussed. Firstly, only matches of the most recent Olympic Games were examined with future analyses needed to examine the robustness of the current models for match success within major international competitions including the Olympic Games. Additionally, matches within all rounds of the competition (regular and playoff) were examined within the current analyses. While a previous study indicated varying match success reliance on team performance indicators within different stages of seasonal competition [26], we expected this to be of little impact for a short-term tournament like the Olympic Games where each success had a substantial impact on final tournament success. Furthermore, analyses were conducted without examination of the impact of prior matches. Previously, accumulated and moderate fatigue from consecutive matches during a Spanish Basketball Federation tournament was suggested to impact three-point shooting accuracy and/or defensive actions [12]. Future examination of the impact of consecutive matches on team performance indicators and match success may clarify the role of fatigue and relevance of physical conditioning for Olympic success. Further, examination of athlete workloads during matches, possibly via wearable technology, in conjunction with team match performance indicators may identify successful team profiles to assist coaches with strategic planning during elite basketball competition.

5. Conclusions

The current study has identified shooting proficiency and defensive actions (e.g., "defensive rebounds", "steals") as quintessential for match success during a women's Olympic basketball tournament. The unique combination of these performance indicators can provide coaches with a greater probability of winning elite matches (>91%). The development of key defensive strategies and/or the selection of athletes highly proficient in, and/or possessing fitness characteristics conducive to, defensive actions may strengthen Olympic match success. The use of non-linear, analytical techniques may provide sport scientists and coaches with superior and practical approaches to exploring multivariate datasets in elite sports for elite success.

Acknowledgments: No financial support was required or provided for this study.

Author Contributions: A.S.L. and C.T.W. conceived and designed the study; A.S.L. and C.T.W. analyzed the data; A.S.L., M.A.G., and C.T.W. wrote, reviewed, and approved the manuscript.

Conflicts of Interest: The authors declare no conflict of interest.

References

1. Fédération Internationale de Basketball (FIBA). Facts & Figures. Available online: http://www.fiba. basketball/presentation#\T1\textbar{}tab=element_2_1 (accessed on 2 October 2017).
2. Scanlan, A.T.; Tucker, P.S.; Dascombe, B.J.; Berkelmans, D.M.; Hiskens, M.I.; Dalbo, V.J. Fluctuations in activity demands across game quarters in professional and semiprofessional male basketball. *J. Strength Cond. Res.* **2015**, *29*, 3006–3015. [CrossRef] [PubMed]
3. Scanlan, A.T.; Dascombe, B.J.; Reaburn, P.; Dalbo, V.J. The physiological and activity demands experienced by Australian female basketball players during competition. *J. Sci. Med. Sport* **2012**, *15*, 341–347. [CrossRef] [PubMed]
4. Klusemann, M.J.; Pyne, D.B.; Hopkins, W.G.; Drinkwater, E.J. Activity profiles and demands of seasonal and tournament basketball competition. *Int. J. Sports Physiol. Perform.* **2013**, *8*, 623–629. [CrossRef] [PubMed]
5. Koklu, Y.; Alemdaroglu, U.; Kocak, F.U.; Erol, A.E.; Findikoglu, G. Comparison of chosen physical fitness characteristics of Turkish professional basketball players by division and playing position. *J. Hum. Kinet.* **2011**, *30*, 99–106. [CrossRef] [PubMed]

6.   Fort-Vanmeerhaeghe, A.; Montalvo, A.; Latinjak, A.; Unnithan, V. Physical characteristics of elite adolescent female basketball players and their relationship to match performance. *J. Hum. Kinet.* **2016**, *53*, 167–178. [CrossRef] [PubMed]

7.   Gomez, M.A.; Lorenzo, A.; Ortega, E.; Sampaio, J.; Ibanez, S.J. Game related statistics discriminating between starters and nonstarters players in Women's National Basketball Association League (WNBA). *J. Sports Sci. Med.* **2009**, *8*, 278–283. [PubMed]

8.   Leicht, A.S.; Gómez, M.A.; Woods, C.T. Explaining match outcome during the Men's basketball tournament at the Olympic games. *J. Sports Sci. Med.* **2017**, *16*, 468–473. [PubMed]

9.   Gomez, M.A.; Lorenzo, A.; Sampaio, J.; Ibanez, S.J.; Ortega, E. Game-related statistics that discriminated winning and losing teams from the Spanish men's professional basketball teams. *Coll. Antropol.* **2008**, *32*, 451–456.

10.   Sampaio, J.; Ibanez, S.; Lorenzo, A.; Gomez, M. Discriminative game-related statistics between basketball starters and nonstarters when related to team quality and game outcome. *Percept. Mot. Skills* **2006**, *103*, 486–494. [CrossRef] [PubMed]

11.   Jukic, I.; Milanovic, D.; Vuleta, D.; Bracic, M. Evaluation of variables of shooting for a goal recorded during the 1997 European Basketball Championship in Barcelona. *Kinesiology* **2000**, *32*, 51–62.

12.   Ibanez, S.J.; Garcia, J.; Feu, S.; Lorenzo, A.; Sampaio, J. Effects of consecutive basketball games on the game-related statistics that discriminate winner and losing teams. *J. Sports Sci. Med.* **2009**, *8*, 458–462. [PubMed]

13.   Lorenzo, A.; Gomez, M.A.; Ortega, E.; Ibanez, S.J.; Sampaio, J. Game related statistics which discriminate between winning and losing under-16 male basketball games. *J. Sports Sci. Med.* **2010**, *9*, 664–668. [PubMed]

14.   Trninic, S.; Dizdar, D.; Luksic, E. Differences between winning and defeated top quality basketball teams in final tournaments of European club championship. *Coll. Antropol.* **2002**, *26*, 521–531. [PubMed]

15.   Gómez, M.A.; Ibáñez, S.J.; Parejo, I.; Furley, P. The use of classification and regression tree when classifying winning and losing basketball teams. *Kinesiology* **2017**, *49*, 47–56.

16.   Robertson, S.; Woods, C.; Gastin, P. Predicting higher selection in elite junior Australian Rules football: The influence of physical performance and anthropometric attributes. *J. Sci. Med. Sport* **2015**, *18*, 601–606. [CrossRef] [PubMed]

17.   Woods, C.T.; Sinclair, W.; Robertson, S. Explaining match outcome and ladder position in the National Rugby League using team performance indicators. *J. Sci. Med. Sport* **2017**, *20*, 1107–1111. [CrossRef] [PubMed]

18.   Sampaio, J.; Godoy, S.I.; Feu, S. Discriminative power of basketball game-related statistics by level of competition and sex. *Percept. Mot. Skills* **2004**, *99*, 1231–1238. [CrossRef] [PubMed]

19.   Gómez, M.A.; Lorenzo, A.; Sampaio, J.; Ibáñez, S.J. Differences between winning and losing teams in women's Basketball game-related statistics. *J. Hum. Move. Stud.* **2006**, *51*, 357–369.

20.   Gomez, M.A.; Perez, J.; Molik, B.; Szyman, R.J.; Sampaio, J. Performance analysis of elite men's and women's wheelchair basketball teams. *J. Sports Sci.* **2014**, *32*, 1066–1075. [CrossRef] [PubMed]

21.   Drinkwater, E.J.; Hopkins, W.G.; McKenna, M.J.; Hunt, P.H.; Pyne, D.B. Modelling age and secular differences in fitness between basketball players. *J. Sports Sci.* **2007**, *25*, 869–878. [CrossRef] [PubMed]

22.   Gomez, M.A.; Lorenzo, A.; Ibanez, S.J.; Sampaio, J. Ball possession effectiveness in men's and women's elite basketball according to situational variables in different game periods. *J. Sports Sci.* **2013**, *31*, 1578–1587. [CrossRef] [PubMed]

23.   Cohen, J. A power primer. *Psychol. Bull.* **1992**, *112*, 155–159. [CrossRef] [PubMed]

24.   Burnham, K.P.; Anderson, D.R. *Model Selection and Multimodel Inference: A Practical Information-Theoretic Approach*, 2nd ed.; Springer-Verlag: New York, NY, USA, 2002.

25.   Hothorn, T.; Hornik, K.; Zeileis, A. Unbiased recursive partitioning: A conditional inference framework. *J. Comput. Graph. Stat.* **2006**, *15*, 651–674. [CrossRef]

26.   Garcia, J.; Ibanez, S.J.; De Santos, R.M.; Leite, N.; Sampaio, J. Identifying basketball performance indicators in regular season and playoff games. *J. Hum. Kinet.* **2013**, *36*, 161–168. [CrossRef] [PubMed]

27.   Csapo, P.; Avugos, S.; Raab, M.; Bar-Eli, M. How should "hot" players in basketball be defended? The use of fast-and-frugal heuristics by basketball coaches and players in response to streakiness. *J. Sports Sci.* **2015**, *33*, 1580–1588. [CrossRef] [PubMed]

28. Gómez, M.A.; Evangelos, T.; Alberto, L. Defensive systems in basketball ball possessions. *Int. J. Perform. Anal. Sport* **2006**, *6*, 98–107.
29. Torres-Ronda, L.; Ric, A.; Llabres-Torres, I.; de las Heras, B.; Schelling i del Alcazar, X. Position-dependent cardiovascular response and time-motion analysis during training drills and friendly matches in elite male basketball players. *J. Strength Cond. Res.* **2016**, *30*, 60–70. [CrossRef] [PubMed]

© 2017 by the authors. Licensee MDPI, Basel, Switzerland. This article is an open access article distributed under the terms and conditions of the Creative Commons Attribution (CC BY) license (http://creativecommons.org/licenses/by/4.0/).

Article

The Effect of Match Schedule on Accelerometry-Derived Exercise Dose during Training Sessions throughout a Competitive Basketball Season

Craig Staunton [1], Daniel Wundersitz [1], Brett Gordon [1], Edhem Custovic [2], Jonathan Stanger [2] and Michael Kingsley [1,*]

[1] Exercise Physiology, La Trobe Rural Health School, La Trobe University, Bundoora 3086, Australia; C.Staunton@latrobe.edu.au (C.S.); D.Wundersitz@latrobe.edu.au (D.W.); B.Gordon@latrobe.edu.au (B.G.)
[2] Computer and Mathematical Sciences, La Trobe University, Bundoora 3086, Australia; E.Custovic@latrobe.edu.au (E.C.); J.Stanger@latrobe.edu.au (J.S.)
* Correspondence: M.Kingsley@latrobe.edu.au; Tel.: +61-3-5444-7589

Received: 25 June 2018; Accepted: 12 July 2018; Published: 23 July 2018

Abstract: Accelerometry-derived exercise dose (intensity $\times$ duration) was assessed throughout a competitive basketball season. Nine elite basketballers wore accelerometers during a Yo-Yo intermittent recovery test (Yo-Yo-IR1) and during three two-week blocks of training that represented phases of the season defined as easy, medium, and hard based on difficulty of match schedule. Exercise dose was determined using accumulated impulse (accelerometry-derived average net force $\times$ duration). Relative exercise intensity was quantified using linear relationships between average net force and oxygen consumption during the Yo-Yo-IR1. Time spent in different intensity zones was computed. Influences of match schedule difficulty and playing position were evaluated. Exercise dose reduced for recovery and pre-match tapering sessions during the medium match schedule. Exercise dose did not vary during the hard match schedule. Exercise dose was not different between playing positions. The majority of activity during training was spent performing sedentary behaviour or very light intensity activity ($64.3 \pm 6.1\%$). Front-court players performed a greater proportion of very light intensity activity (mean difference: $6.8 \pm 2.8\%$), whereas back-court players performed more supramaximal intensity activity (mean difference: $4.5 \pm 1.0\%$). No positional differences existed in the proportion of time in all other intensity zones. Objective evaluation of exercise dose might allow coaches to better prescribe and monitor the demands of basketball training.

Keywords: female; training load; monitoring; accelerometer; workloads

1. Introduction

Training sessions contribute substantially to the total volume of exercise that basketball players receive (exercise dose; product of exercise intensity and duration) during the competitive season [1]. Although the exercise dose during basketball match-play has been extensively examined [2–7], the exercise dose associated with training sessions remains largely unreported [8]. Only one study to date has investigated the exercise dose received by players during the in-season phase of a basketball training program [1]. The results from this study showed that match schedule (i.e., no match, one match, or two matches per week) influences the exercise dose received by players [1]. However, these data were collected from only one two-week block during the in-season phase of competition. Previous research has identified fluctuations in exercise intensity during different phases of basketball pre-season training [8]. Therefore, it is plausible that the exercise dose received by players fluctuates throughout different phases of a competitive basketball season. However, no research to date has investigated the exercise dose received by players throughout multiple phases of a competitive basketball season.

The exercise dose and intensity of activity during basketball match-play and training are most commonly quantified using time-motion analyses and physiological data [2,3,6,9–12]. For example, movement speeds (derived from time-motion analyses) and heart rate responses are often used to measure exercise intensity. While time-motion analyses can be used to assess movement patterns undertaken and physiological responses provide a measure of average exercise intensity, the highly intermittent pattern of exercise and frequent vertical efforts make these methods inappropriate to quantify exercise dose during basketball match-play and training. In support of this statement, time-motion analyses underestimate the external demands of basketball-specific movements (e.g., jumping, shuffling, changes of direction) [13] and physiological analyses are associated with delays in responsiveness due to cardiorespiratory lag [14]. Thus, these techniques are incapable of accurately quantifying brief bouts of supramaximal intensity exercise and rapid changes in movements that occur frequently in basketball [2,3,6].

With the aim to circumvent the aforementioned limitations that are associated with other measurement systems, wearable accelerometers have emerged as an alternative method to quantify exercise dose during basketball. Accelerometers have high data acquisition rates and can measure activity in three planes of motion, making this measurement technique well-suited to quantifying the exercise dose and intensity in intermittent sports, such as basketball.

Average net force (AvF_{Net}) is an accelerometry-derived measure of exercise intensity, with confirmed construct validity in basketball [13]. Strong relationships between accelerometry-derived metrics and oxygen consumption ($\dot{V}O_2$) have been previously identified [7,15,16], exemplifying that accelerometers can be used to estimate relative exercise intensity. Additionally, supramaximal intensity exercise can be estimated from extrapolation of individual linear relationships between running speed and oxygen consumption [17–19]. Therefore, AvF_{Net} offers a measurement technique that is well-suited to calculate relative exercise intensity during intermittent sports, including the measurement of supramaximal intensity efforts. In addition to quantifying exercise intensity, the product of AvF_{Net} and exercise duration (Impulse) can be used to quantify exercise dose. Consequently, accelerometery-derived AvF_{Net} can provide a suitable method to quantify the relative exercise intensity completed by players during basketball training sessions, which could help coaches to prescribe more match-specific training and execute periodised training plans.

The aim of this study was to use accelerometry-derived AvF_{Net} and accumulated impulse to assess relative exercise intensity and exercise dose during training sessions completed at different phases of a competitive basketball season.

2. Materials and Methods

2.1. Participants

Nine professional players (27 ± 5 years, 182 ± 8 cm, 81 ± 12 kg) from a basketball team competing in the Australian Women's National Basketball League (WNBL) participated in this study. All players provided informed written consent and completed the requirements of this study. Ethical approval was granted by the La Trobe University Human Research Ethics Committee (ref: UHEC 15-088).

2.2. Study Design

All players completed preliminary testing and were monitored over the course of a 17-round competitive basketball season. Six separate weeks of training data were collected during the competitive basketball season from three separate phases. Two weeks of training were monitored from each phase, where each week consisted of three team training sessions. The phases of monitored training were selected to represent periods of different match schedule difficulty, defined as easy, medium, and hard. The easy match schedule occurred between rounds 10 to 12, where the team played home matches against the two lowest ranked teams in the competition. The medium match schedule occurred between rounds 7 to 8, where the team played one double-header (i.e., two matches within a single round of competition) and one away match against moderately positioned teams. The hard match schedule occurred between rounds 4 to 6, where the team had an extensive travel schedule (away double-header) and a home match against the competition grand finalist (Figure 1).

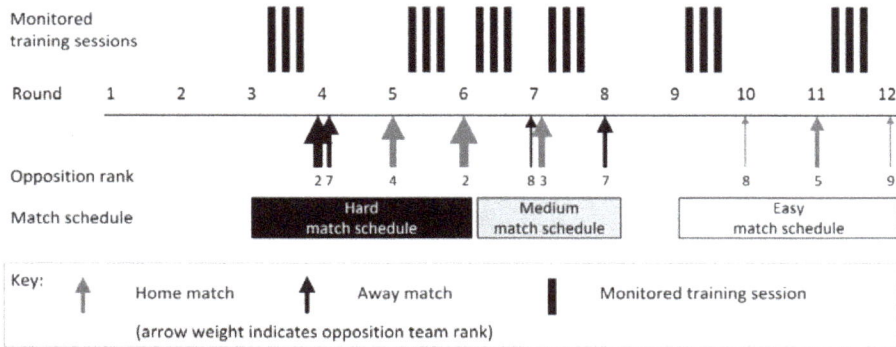

Figure 1. Schematic representation of data collection across the basketball season.

Preliminary testing included the measurement of body mass and standing stature according to the International Society for the Advancement of Kinanthropometry guidelines and procedures [20]. Additionally, a modified Yo-Yo intermittent recovery test (level 1; Yo-Yo-IR1), which included seven additional low speed stages prior to the commencement of the Yo-Yo-IR1, was completed. Movement speeds started at 3 km·h^{-1} and progressed by 1 km·h^{-1} for each stage until 9 km·h^{-1}; after this, the original Yo-Yo-IR1 test was completed until exhaustion. Slower movement speeds occur frequently throughout basketball match-play [2,6], and inclusion of the slower movement speeds allowed for the calibration of relative exercise intensity across a broad range of movement speeds.

During all physical testing and training sessions the players wore a commercially available tri-axial accelerometer (Link; ActiGraph, Pensacola, FL, USA) on the upper-back as previously described [7,13], which recorded accelerations at 100 Hz. Previous research has established high levels of reliability for ActiGraph accelerometers [21–23]. In addition, breath-by-breath oxygen consumption (Oxycon Mobile, Jaeger, Germany) was recorded during the modified Yo-Yo-IR1 in order to establish individual relationships between accelerometry and $\dot{V}O_2$.

2.3. Data Analyses

Accelerometer data were downloaded using the manufacturer's software (ActiLife v12; ActiGraph, USA). Exercise intensity was quantified using AvF$_{Net}$ as previously described [7,13]. To calculate AvF$_{Net}$, the three planes of tri-axial accelerations were filtered using a dual-pass, fourth order Butterworth filter (high pass: 0.1 Hz, low pass: 15 Hz). These cut-off frequencies were chosen to remove gravity [24,25] and noise [26,27] components, respectively. After filtering, the product of the instantaneous acceleration vector and player's body mass was used to determine instantaneous net force (F$_{Net}$). The average F$_{Net}$ (AvF$_{Net}$) for user-selected periods was calculated in 1-s epochs using customised software (LabVIEW 2016; National Instruments, Austin, Texas, USA). In addition, interpolated $\dot{V}O_2$ was included in the output from the modified Yo-Yo-IR1. To quantify the exercise dose for the entire training session, the numerical integral of AvF$_{Net}$ and exercise duration was used to calculate accumulated impulse (Impulse), measured in Newton seconds (N·s).

Resting $\dot{V}O_2$ was determined during 5-min seated rest prior to the beginning of the modified Yo-Yo-IR1. Accelerometer and $\dot{V}O_2$ data were synchronised during the modified Yo-Yo-IR1 during the initial shuttle, where the acceleration signal was reconciled with the commencement of $\dot{V}O_2$ recording. For all stages of the modified Yo-Yo-IR1 the acceleration signal was selected from the commencement of movement, which was identified as the moment when the resultant acceleration began to rise from rest, until the completion of the 40-m stage using the custom software to determine AvF_{Net} and $\dot{V}O_2$. Peak $\dot{V}O_2$ was determined as the greatest 5-s average $\dot{V}O_2$ achieved during the final completed Yo-Yo-IR1 stage. $\dot{V}O_2$ reserve ($\dot{V}O_2R$) was calculated for each individual in order to represent relative maximum $\dot{V}O_2$ above rest, by subtracting resting $\dot{V}O_2$ from peak $\dot{V}O_2$. Subsequently, AvF_{Net} and average $\dot{V}O_2R$ for each completed stage were correlated and best-fit linear relationships were generated for all players ($r^2 = 0.93$–0.97).

Accelerometry data from all recorded training sessions included all activity, stoppages, and time-outs beginning from the commencement of the warm-up to the completion of the final drill or cool-down. Training schedules included three separate training sessions per week (Sessions 1–3) with each training session consisting of warm-up drills, skill drills, offensive and defensive technical/tactical drills, and match-simulation drills.

Predicted $\dot{V}O_2R$ during training sessions were determined from AvF_{Net} (1-s epochs) using the player's linear relationship developed from the Yo-Yo-IR1. Relative exercise intensity was categorised into seven intensity zones similar to those identified by the American College of Sports Medicine [28] being: sedentary behaviour ($<20\%$ $\dot{V}O_2R$); very light (20–$<30\%$ $\dot{V}O_2R$); light (30–$<40\%$ $\dot{V}O_2R$); moderate (40–$<60\%$ $\dot{V}O_2R$); vigorous (60–$<90\%$ $\dot{V}O_2R$); maximal (90–$<100\%$ $\dot{V}O_2R$); and supramaximal ($\geq100\%$ $\dot{V}O_2R$). Total time and proportion of time in all intensity zones were determined for all players across all training sessions. Outcome measures were calculated for all players and data were separated by playing position: front-court players (small forwards, power forwards, and centres; $n = 5$) and back-court players (point guards and shooting guards; $n = 4$).

2.4. Statistical Analyses

Statistical analyses were completed using IBM SPSS Statistics (v24; IBM Corporation, Armonk, NY, USA). Shapiro-Wilk tests confirmed that the assumption of normality was not violated, and group data were expressed as mean $\pm$ standard deviation (SD). Repeated measures two-way mixed model analyses of variance (ANOVA) (within factors: Match schedule and Session; between factor: Position) was used to determine the effect of match schedule difficulty, session, and position on exercise dose (Impulse) and intensity (AvF_{Net} and the proportion of time in all intensity zones). Effect sizes are presented as partial eta-squared statistic (η^2_p). Mauchly's test was consulted and Greenhouse–Geisser correction was applied if the assumption of sphericity was violated. Significant interactions or main effects were followed up with simple main effect analyses with pairwise comparisons using Bonferroni correction. Significance was set at $p < 0.05$.

3. Results

Average exercise intensity (AvF_{Net}) across all 18 training sessions was 293 ± 40 N and was not different between match schedule, session, or playing position. The majority of activity during training was spent performing sedentary behaviour or very light intensity activity ($64.3 \pm 6.1\%$). Front-court position players performed a greater proportion of very light intensity activity during training sessions when compared with back-court players (mean difference: $6.8 \pm 2.8\%$; Position effect: $F_{(1,7)} = 5.798$; $p = 0.047$; $\eta^2_p = 0.453$). Back-court position players performed more supramaximal intensity activity when compared with front-court players during the medium match schedule (mean difference: $4.5 \pm 1.0\%$; Match schedule $\times$ Position interaction: $F_{(2,14)} = 9.323$; $p = 0.003$; $\eta^2_p = 0.573$). There were no positional differences in the proportion of time in all other intensity zones across all three match schedules (Table 1).

Table 1. Proportion of total duration (%) spent in each intensity zone for front-court and back-court players for each phase of the competitive season.

	Sedentary	Very Light	Light	Moderate	Vigorous	Maximal	Supra-Maximal
Easy							
front-court	38.5 ± 10.0	22.3 ± 5.6	12.1 ± 4.7	11.0 ± 3.3	11.5 ± 2.7	2.3 ± 1.0	2.2 ± 1.4
back-court	45.9 ± 7.6	16.4 ± 1.4	9.4 ± 2.9	10.2 ± 4.4	11.2 ± 1.2	3.3 ± 1.4	3.6 ± 1.7
Medium							
front-court	40.7 ± 13.1	21.6 ± 5.8	11.9 ± 4.9	10.3 ± 3.7	11.6 ± 3.8	2.4 ± 0.6	1.4 ± 0.5 *
back-court	52.2 ± 4.1	14.4 ± 1.0	7.2 ± 1.6	6.7 ± 1.6	9.9 ± 4.1	3.6 ± 1.4	5.9 ± 2.2
Hard							
front-court	43.8 ± 10.7	20.8 ± 5.2	10.9 ± 5.6	8.6 ± 2.4	12.0 ± 3.4	2.5 ± 0.9	1.3 ± 0.5
back-court	51.8 ± 6.4	14.4 ± 1.0	6.7 ± 2.0	6.8 ± 1.6	10.2 ± 4.7	3.5 ± 1.5	7.0 ± 5.8
Total							
front-court	40.8 ± 10.9	21.6 ± 5.4 *	11.7 ± 4.8	10.1 ± 2.8	11.7 ± 3.0	2.4 ± 0.6	1.7 ± 0.7
back-court	50.2 ± 4.4	14.8 ± 6.7	7.8 ± 1.8	7.8 ± 2.0	10.3 ± 3.2	3.5 ± 0.9	5.6 ± 2.7

Mean $\pm$ standard deviation. * Different to back-court ($p < 0.05$). $\dot{V}O_2R$: Volume of oxygen uptake reserve. Sedentary: $<20\%$ $\dot{V}O_2R$; Very Light: $20–<30\%$ $\dot{V}O_2R$; Light: $30–<40\%$ $\dot{V}O_2R$; Moderate: $40–<60\%$ $\dot{V}O_2R$; Vigorous: $60–<90\%$ $\dot{V}O_2R$; Maximal: $90–<100\%$ $\dot{V}O_2R$; Supramaximal: $\geq 100\%$ $\dot{V}O_2R$.

The proportion of time performing very light intensity activity was different according to difficulty of match schedule (Match schedule effect: $F_{(2,14)} = 4.761$; $p = 0.026$; $\eta^2_p = 0.405$), where more very light intensity activity tended to be performed during the easy match schedule compared with the hard match schedule (mean difference: $2.0 \pm 2.0\%$; $p = 0.062$). Match schedule difficultly had no influence on the proportion of time in all other intensity zones. The proportion of vigorous intensity activity was different between sessions (Session effect: $F_{(2,14)} = 5.271$; $p = 0.020$; $\eta^2_p = 0.430$). There was a greater proportion of vigorous intensity activity during Session 3 when compared with Session 1 (mean difference: $1.5 \pm 1.2\%$; $p = 0.026$) through each match schedule.

Mean exercise dose (Impulse) across all 18 training sessions was 1939 ± 258 kN·s. Playing position had no influence on the exercise dose received across match schedules (Position $\times$ Match schedule interaction: $F_{(2,14)} = 0.133$; $p = 0.877$; $\eta^2_p = 0.019$) or training sessions (Position $\times$ Session interaction: $F_{(2,14)} = 0.374$; $p = 0.695$; $\eta^2_p = 0.051$). The pattern of exercise dose during the three team training sessions per week changed according to the difficulty of match schedule (Figure 2; Match schedule x Session interaction: $F_{(4,28)} = 4.224$; $p = 0.008$; $\eta^2_p = 0.376$). Exercise dose during Session 2 was greater when compared with Session 1 (mean difference: 537 ± 156 kN·s; $p = 0.010$) and Session 3 (mean difference: 476 ± 186 kN·s; $p = 0.001$) during the medium match schedule and greater compared with Session 1 (mean difference: 509 ± 107 kN·s; $p = 0.01$) during the easy match schedule. Exercise dose was similar between sessions during the hard match schedule ($p \geq 0.941$).

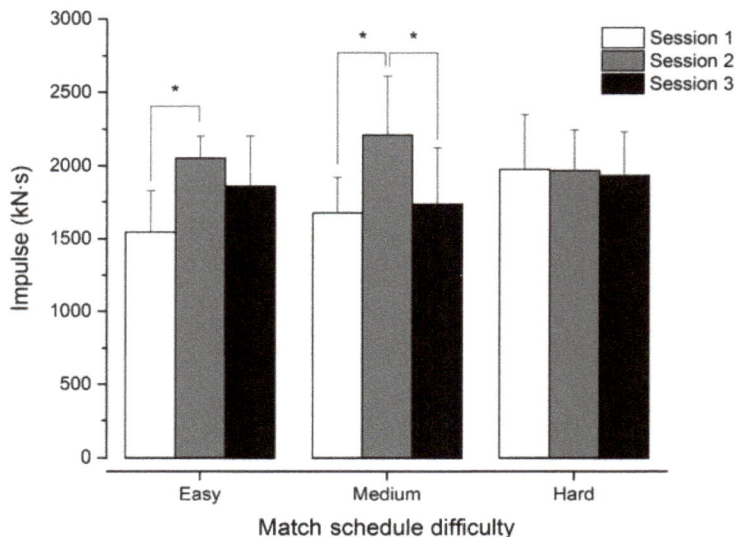

Figure 2. Exercise dose (Impulse) for easy, medium, and hard match schedules. Mean ± standard deviation. * Different between sessions ($p < 0.05$).

4. Discussion

This is the first study to assess the relative exercise intensity of basketball training using a method that is suitable for sports involving rapid changes in movement patterns and intensities. The main findings of this study demonstrate that exercise dose varied between training sessions (i.e., Session 1, Session 2, Session 3) during easy and moderate difficulty match schedules but not during hard match schedules. Match schedule had no influence on the average exercise intensity and limited influence on the proportion of time spent in each intensity zone. Furthermore, few position-specific differences existed in the exercise dose, average exercise intensity, or the proportion of time spent in each intensity zone during training sessions completed by an elite women's basketball team.

The present study identified that the majority of exercise during basketball training sessions (64%) was spent performing either sedentary behaviour or very light intensity exercise. Results from a previous study, which used similar methods to the current study, show that a slightly lower proportion (approximately 59%) of match-play was spent performing either sedentary behaviour or very light intensity exercise [7]. Additionally, previous time-motion analyses have identified lower proportions (30–42%) of live match-play performing low-intensity and recovery activities (e.g., standing, walking) [6,9]. Taken together, these findings suggest that basketball training sessions are associated with greater periods of sedentary behaviour, likely due to inclusion of technical/tactical drills that involve large portions of time standing and walking while receiving coaching instruction [29]. Coaches should be aware that providing large amounts of instruction might compromise the match-specificity of training sessions. Therefore, technical/tactical drills can be combined with conditioning goals in order to more closely replicate match demands [30].

Despite large proportions of sedentary behaviour and very light intensity exercise during basketball, previous investigations consistently report high average physiological responses over the course of basketball training sessions. For example, mean $\dot{V}O_2$ values during basketball training have been reported in the range of 60–80% $\dot{V}O_{2max}$ [4,31]. Additionally, heart rate responses are typically in the range of 85–90% of maximum [31–34]. High physiological intensities during basketball have been reported because brief rest periods and active recovery (i.e., walking and jogging) during basketball are insufficient to permit full physiological recovery [12]. Thus, accelerated $\dot{V}O_2$ kinetics at the onset of a work interval, in addition to cardiorespiratory lag, means that the physiological response at a particular point in time does not directly reflect the actual intensity of activity being undertaken. Therefore, physiological responses such as heart rate and $\dot{V}O_2$ cannot truly reflect the exercise intensity during intermittent exercise, such as basketball. This highlights the importance of using accelerometry-derived AvF_{Net} to accurately quantify short duration bouts of intermittent exercise. This method could be important for athlete monitoring in order for coaches to replicate the most demanding aspects of match-play in training sessions, which can maximise training benefits and improve performance [35]. Additionally, findings from previous research suggest that periods of higher exercise dose throughout a basketball season are associated with greater risk of injury [36]. Therefore, monitoring accelerometry-derived AvF_{Net} throughout a basketball season might be useful to identify periods of heightened injury risk as a consequence of elevated exercise dose.

The present study identified that basketball training sessions elicit a similar accelerometry-derived exercise dose and proportion of time in each relative exercise intensity zone between front-court and back-court playing positions. This finding corroborates previous research, which found that movement demands during basketball training, assessed via time-motion analyses, were largely similar between playing positions [29]. Conversely, these findings are in direct contrast to the positional differences observed via time-motion analyses and physiological responses during basketball match-play [2,6,9,12]. Furthermore, recent evidence identified that these positional differences extend to accelerometry-derived relative exercise intensities during basketball match-play [7], suggesting that exercise dose and proportion of time in each exercise intensity zone during training is not always reflective of match-play. Similar exercise dose and intensity between playing positions during training sessions indicates that the individual positional demands of match-play are not replicated during training sessions. This might be due to logistical factors, such as a lack of time, space, and resources, which can make it difficult for coaches to individualise training for team sports. As such, during team training sessions, all players are often prescribed the same training drills. For this study, no feedback regarding players' exercise dose or proportion of time in each intensity zone was provided to the coach, thus the exercise dose received is based solely upon the coach's exercise prescription and the exercise completed by players. Therefore, it is possible that providing objective feedback of the exercise dose received by players and proportion of time in each intensity zone could assist coaching staff to calibrate exercise prescriptions and better replicate the exercise dose from match-play.

Both the exercise dose received by athletes and proportion of time in each intensity zone remained largely similar across the course of the season, despite variability in the difficulty of match schedule. Nevertheless, exercise dose varied between training sessions (i.e., Session 1, Session 2, Session 3) during the easy and medium match schedules. Increased exercise dose during Session 2 during the easy and moderate match schedules might be the coach's attempt to compensate for the reduced exercise dose during the recovery session (Session 1) and pre-match tapering (Session 3). On the other hand, exercise dose did not vary between training sessions during the hard match schedule. These findings corroborate previous research from professional men's basketball, which identified that exercise dose is related to the competition schedule [1]. Specifically, subjectively measured exercise dose (rating of perceived exertion × duration) from training sessions was also lower both pre- and post-match [1]. It is well-established that tapering can assist in improving competition performance [37]; however, future research should assess the most effective tapering strategy for the unique demands of basketball competition that often involve one or two competitive matches every week.

5. Conclusions

The exercise dose and intensity received by athletes remained largely similar throughout the competitive season despite variability in the difficulty of match schedule. Although coaches might be reducing exercise dose for recovery and pre-match tapering during easy and moderate difficulty match schedules, there was no evidence of training periodisation during hard match schedule. Furthermore, there were few position-specific differences in exercise dose and proportion of time in each intensity zone over the course of an elite women's basketball season. Objective monitoring of the exercise dose in training and match-play via accelerometry-derived AvF$_{Net}$ might enable coaches to better prescribe match-specific exercise during training.

Author Contributions: M.K. and C.S. conceived and designed the experiments; C.S. performed the experiments; M.K. developed the customised LabVIEW program; C.S. and J.S. analysed the data with assistance from E.C., M.K., J.S., D.W., and B.G.; C.S. wrote the paper with editorial assistance from M.K., B.G., D.W., and E.C.

Funding: This research received no external funding.

Acknowledgments: This work was supported by an Australian Government Research Training Program Scholarship. The authors would like to thank the Bendigo Spirit basketball players and staff for their assistance and commitment to this research.

Conflicts of Interest: The authors declare no conflict of interest.

References

1. Manzi, V.; D'Ottavio, S.; Impellizzeri, F.M.; Chaouachi, A.; Chamari, K.; Castagna, C. Profile of weekly training load in elite male professional basketball players. *J. Strength Cond. Res.* **2010**, *24*, 1399–1406. [CrossRef] [PubMed]
2. Abdelkrim, N.B.; El Fazaa, S.; El Ati, J. Time–motion analysis and physiological data of elite under-19-year-old basketball players during competition. *Br. J. Sports Med.* **2007**, *41*, 69–75. [CrossRef] [PubMed]
3. Conte, D.; Favero, T.G.; Lupo, C.; Francioni, F.M.; Capranica, L.; Tessitore, A. Time-motion analysis of Italian elite women's basketball games: Individual and team analyses. *J. Strength Cond. Res.* **2015**, *29*, 144–150. [CrossRef] [PubMed]
4. Montgomery, P.G.; Pyne, D.B.; Minahan, C.L. The physical and physiological demands of basketball training and competition. *Int. J. Sports Physiol. Perform.* **2010**, *5*, 75–86. [PubMed]
5. Narazaki, K.; Berg, K.; Stergiou, N.; Chen, B. Physiological demands of competitive basketball. *Scand. J. Med. Sci. Sports* **2009**, *19*, 425–432. [CrossRef] [PubMed]
6. Scanlan, A.; Dascombe, B.; Reaburn, P. A comparison of the activity demands of elite and sub-elite Australian men's basketball competition. *J. Sports Sci.* **2011**, *29*, 1153–1160. [CrossRef] [PubMed]
7. Staunton, C.; Wundersitz, D.; Gordon, B.; Kingsley, M. Accelerometry-derived relative exercise intensities in elite women's basketball. *Int. J. Sports Med.* **2018**. [CrossRef] [PubMed]
8. Fox, J.L.; Stanton, R.; Scanlan, A. A Comparison of Training and Competition Demands in Semiprofessional Male Basketball Players. *Res. Q. Exerc. Sport* **2018**, *89*, 103–111. [CrossRef] [PubMed]
9. Scanlan, A.; Dascombe, B.J.; Reaburn, P.; Dalbo, V.J. The physiological and activity demands experienced by Australian female basketball players during competition. *J. Sci. Med. Sport* **2012**, *15*, 341–347. [CrossRef] [PubMed]
10. Rodriguez-Alonso, M.; Fernandez-Garcia, B.; Perez-Landaluce, J.; Terrados, N. Blood lactate and heart rate during national and international women's basketball. *J. Sports Med. Phys. Fit.* **2003**, *43*, 432–436.
11. Moreira, A.; McGuigan, M.R.; Arruda, A.F.; Freitas, C.G.; Aoki, M.S. Monitoring internal load parameters during simulated and official basketball matches. *J. Strength Cond. Res.* **2012**, *26*, 861–866. [CrossRef] [PubMed]
12. Stojanović, E.; Stojiljković, N.; Scanlan, A.T.; Dalbo, V.J.; Berkelmans, D.M.; Milanović, Z. The Activity Demands and Physiological Responses Encountered During Basketball Match-Play: A Systematic Review. *Sports Med.* **2018**, *48*, 111–135. [CrossRef] [PubMed]
13. Staunton, C.; Wundersitz, D.; Gordon, B.; Kingsley, M. Construct validity of accelerometry-derived force to quantify basketball movement patterns. *Int. J. Sports Med.* **2017**, *38*, 1090–1096. [CrossRef] [PubMed]
14. Xu, F.; Rhodes, E.C. Oxygen uptake kinetics during exercise. *Sports Med.* **1999**, *27*, 313–327. [CrossRef] [PubMed]
15. Fudge, B.W.; Wilson, J.; Easton, C.; Irwin, L.; Clark, J.; Haddow, O.; Kayser, B.; Pitsiladis, Y.P. Estimation of oxygen uptake during fast running using accelerometry and heart rate. *Med. Sci. Sports Exerc.* **2007**, *39*, 192–198. [CrossRef] [PubMed]
16. McGregor, S.J.; Busa, M.A.; Yaggie, J.A.; Bollt, E.M. High resolution MEMS accelerometers to estimate VO$_2$ and compare running mechanics between highly trained inter-collegiate and untrained runners. *PLoS ONE* **2009**, *4*, e7355. [CrossRef] [PubMed]

17. Medbo, J.I.; Mohn, A.-C.; Tabata, I.; Bahr, R.; Vaage, O.; Sejersted, O.M. Anaerobic capacity determined by maximal accumulated O2 deficit. *J. Appl. Physiol.* **1988**, *64*, 50–60. [CrossRef] [PubMed]
18. Noordhof, D.A.; De Koning, J.J.; Foster, C. The maximal accumulated oxygen deficit method. *Sports Med.* **2010**, *40*, 285–302. [CrossRef] [PubMed]
19. Scott, C.; Roby, F.; Lohman, T.; Bunt, J. The maximally accumulated oxygen deficit as an indicator of anaerobic capacity. *Med. Sci. Sports Exerc.* **1991**, *23*, 618–624. [CrossRef] [PubMed]
20. Marfell-Jones, M.J.; Stewart, A.; de Ridder, J. *International Standards for Anthropometric Assessment*; ISAK: Potchefstroom, South Africa, 2012.
21. Aadland, E.; Ylvisåker, E. Reliability of the Actigraph GT3X+ accelerometer in adults under free-living conditions. *PLoS ONE* **2015**, *10*, e0134606. [CrossRef] [PubMed]
22. Santos-Lozano, A.; Torres-Luque, G.; Marín, P.J.; Ruiz, J.R.; Lucia, A.; Garatachea, N. Intermonitor variability of GT3X accelerometer. *Int. J. Sports Med.* **2012**, *33*, 994–999. [CrossRef] [PubMed]
23. McClain, J.J.; Sisson, S.B.; Tudor-Locke, C. Actigraph accelerometer interinstrument reliability during free-living in adults. *Med. Sci. Sports Exerc.* **2007**, *39*, 1509–1514. [CrossRef] [PubMed]
24. Boonstra, M.C.; van der Slikke, R.M.; Keijsers, N.L.; van Lummel, R.C.; de Waal Malefijt, M.C.; Verdonschot, N. The accuracy of measuring the kinematics of rising from a chair with accelerometers and gyroscopes. *J. Biomech.* **2006**, *39*, 354–358. [CrossRef] [PubMed]
25. Dalen, T.; Jørgen, I.; Gertjan, E.; Havard, H.G.; Ulrik, W. Player Load, Acceleration, and Deceleration During Forty-Five Competitive Matches of Elite Soccer. *J. Strength Cond. Res.* **2016**, *30*, 351–359. [CrossRef] [PubMed]
26. Wundersitz, D.; Gastin, P.; Robertson, S.; Davey, P.; Netto, K. Validation of a Trunk-mounted Accelerometer to Measure Peak Impacts during Team Sport Movements. *Int. J. Sports Med.* **2015**, *36*, 742–746. [CrossRef] [PubMed]
27. Wundersitz, D.; Gastin, P.B.; Robertson, S.J.; Netto, K.J. Validity of a Trunk Mounted Accelerometer to Measure Physical Collisions in Contact Sports. *Int. J. Sports Physiol. Perform.* **2015**, *10*, 681–686. [CrossRef] [PubMed]
28. Pescatello, L.S.; Arena, R.; Riebe, D.; Thompson, P.D. *ACSM's Guidelines for Exercise Testing and Prescription*, 9th ed.; Wolters Kluwer Health/Lippincott Williams & Wilkins: Philadelphia, PA, USA, 2014.
29. Torres-Ronda, L.; Ric, Á.; de Miguel, B.; llabres, I.; Schelling, X. Position-Dependent Cardiovascular Response and Time-Motion Analysis during Training Drills and Friendly Matches in Elite Male Basketball Players. *J. Strength Cond. Res.* **2016**, *30*, 60–70. [CrossRef] [PubMed]
30. Schelling, X.; Torres-Ronda, L. Conditioning for basketball: Quality and quantity of training. *Strength Cond. J.* **2013**, *35*, 89–94. [CrossRef]
31. Castagna, C.; Impellizzeri, F.M.; Chaouachi, A.; Abdelkrim, B.N.; Manzi, V. Physiological responses to ball-drills in regional level male basketball players. *J. Sports Sci.* **2011**, *29*, 1329–1336. [CrossRef] [PubMed]
32. Klusemann, M.J.; Pyne, D.B.; Foster, C.; Drinkwater, E.J. Optimising technical skills and physical loading in small-sided basketball games. *J. Sports Sci.* **2012**, *30*, 1463–1471. [CrossRef] [PubMed]
33. Scanlan, A.; Wen, N.; Tucker, P.S.; Dalbo, V.J. The Relationships Between Internal and External Training Load Models During Basketball Training. *J. Strength Cond. Res.* **2014**, *28*, 2397. [CrossRef] [PubMed]
34. Sampaio, J.; Abrantes, C.; Leite, N. Power, heart rate and perceived exertion responses to 3x3 and 4x4 basketball small-sided games. *Rev. Psicol. Deporte* **2009**, *18*, 463–467.
35. Gabbett, T.J. GPS analysis of elite women's field hockey training and competition. *J. Strength Cond. Res.* **2010**, *24*, 1321–1324. [CrossRef] [PubMed]
36. Anderson, L.; Triplett-Mcbride, T.; Foster, C.; Doberstein, S.; Brice, G. Impact of training patterns on incidence of illness and injury during a women's collegiate basketball season. *J. Strength Cond. Res.* **2003**, *17*, 734–738. [CrossRef]
37. Bosquet, L.; Montpetit, J.; Arvisais, D.; Mujika, I. Effects of tapering on performance: A meta-analysis. *Med. Sci. Sports Exerc.* **2007**, *39*, 1358–1365. [CrossRef] [PubMed]

© 2018 by the authors. Licensee MDPI, Basel, Switzerland. This article is an open access article distributed under the terms and conditions of the Creative Commons Attribution (CC BY) license (http://creativecommons.org/licenses/by/4.0/).

Article

Scoring Strategies Differentiating between Winning and Losing Teams during FIBA EuroBasket Women 2017

Daniele Conte * and Inga Lukonaitiene

Institute of Sport Science and Innovations, Lithuanian Sports University, 44221 Kaunas, Lithuania;
inga.lukonaitiene@lsu.lt
* Correspondence: daniele.conte@lsu.lt; Tel.: +370-69521927

Received: 30 April 2018; Accepted: 25 May 2018; Published: 29 May 2018

Abstract: This study aimed to examine the scoring strategies differentiating between winning and losing teams during FIBA EuroBasket Women 2017 in relation to different game scores. Data were gathered for all games of FIBA EuroBasket Women 2017 from the official website. The investigated scoring strategies were fast break points (FBP); points in the paint (PP); points from turnover (PT); second chance points (SCP); and points from the bench (PB). Games were classified with cluster analysis based on their score difference as close, balanced, and unbalanced and the differences in the scoring strategies between winning and losing teams were assessed using magnitude-based statistics. Results revealed no substantial differences in FBP in any investigated cluster. Furthermore, winning teams showed a substantially higher number of PP and PT (in close and unbalanced games) and SCP (in balanced and unbalanced games) compared to losing teams. Finally, winning teams scored substantially lower and higher number of BPs in close games and unbalanced games, respectively, compared to losing teams. In conclusion, all the investigated scoring strategies discriminate between winning and losing teams in elite women's basketball except for FBP. These results provide useful information for basketball coaches to optimize their training sessions and game strategies.

Keywords: game-related statistics; performance analysis; basketball performance; team sports; basketball tactics

---

1. Introduction

Basketball is one of the most popular sports worldwide and in particular women's basketball is increasing its popularity [1]. In the last few years, an increasing number of researchers have quantified the performance profile of women's basketball from a physical and physiological standpoint, documenting that women's basketball games are characterized by intermittent high-intensity efforts separated by short recovery periods and a high physiological demand [2,3]. In addition, the technical and tactical performance profile of women's basketball games has been well investigated [4–7]. From a tactical standpoint, previous studies investigated the most effective tactical parameter during ball possessions, documenting that fast break might be one of the main indicators differentiating between winning and losing teams in both women and men's basketball [8,9]. Indeed, it has been demonstrated that winning teams perform a higher number of fast break actions than losing teams [9]. Intuitively, performing more fast break actions would produce more scored points from this action. In addition, further studies documented that the use of the inside game might be considered a fundamental parameter in order to win a basketball game. In this regard, a previous investigation showed that ball possessions including the inside pass were the most effective [10]. However, no previous studies analyzed these indicators in women's basketball. In addition, no studies verified whether the point scored with these tactical strategies (i.e., fast break and inside game actions) might be an indicator able to differentiate between winning and losing teams. In fact, both fast break and inside game strategies might correspond to a higher scored fast break points and points in the paint [10,11]. Therefore, further studies investigating these scoring strategies are warranted.

From a technical standpoint, many studies investigated the game-related statistics differentiating between winning and losing teams in women's basketball [5,7,12]. Previous investigations identified that two of the game-related statistics most discriminating between winning and losing teams are turnovers and rebounds in women's basketball [6,7,12]. Possibly, turnovers provide more opportunities for the opponents to score a basket since the opposing team might steal the ball and run fast break, outnumbering the defense [11]. Similarly, offensive rebounds create a second chance to score for the offensive teams. However, no previous studies investigated whether the points scored from turnover and the second chance points are performance indicators differentiating between winning and losing teams. Therefore, future studies should deeply investigate these aspects. In addition, the bench players' performance can be considered as one of the possible determinants of a win in elite basketball [13]. Previous investigations indicated that bench players might provide a fundamental contribution to win a game, in particular for high-ranked teams [13,14]. Sampaio et al. [13] documented that starter players performed a higher number of defensive rebounds and assists. However, it has been demonstrated that the best teams possibly lose games because of the worse performance of bench players and particularly their offensive performance [13]. Indeed, bench players receive a statistically lower number of fouls and consequently score fewer points from free throws [13]. Therefore, the points scored by bench players might be a discriminant factor differentiating between winning and losing teams. Since it was not previously investigated whether points from the bench might discriminate between winning and losing teams, future studies should address this issue.

The above-mentioned scoring strategies might change in relation to different game scores. Indeed, games with a low or high score difference showed different performance indicators differentiating between winning and losing teams in elite women's basketball [6]. Therefore, the aim of the study was to examine the scoring strategies differentiating between winning and losing teams during FIBA EuroBasket Women 2017 in relation to different game scores.

2. Materials and Methods

2.1. Subjects

The study was approved by an institutional review board, and meets the ethical standards in sports and exercise science research [15]. The game related statistics of all 40 games played in the FIBA EuroBasket Women 2017 were investigated (average score difference: 11.9 ± 8.6 points).

2.2. Procedures

In the tournament, sixteen teams competed in four groups at the preliminary round. Only the top two teams from each group qualified for the final stages (i.e., quarterfinals and final four) competing for the 1st–8th place. Data were gathered from the official box score on the website of the FIBA EuroBasket Women 2017 (http://www.fiba.basketball/eurobasketwomen/2017). The considered game-related statistics referring to scoring strategies were as follows: (a) fast break points (FBP), which refer to the points scored during fast break actions; (b) points in the paint (PP), which indicate the point scored in the key area; (c) points from turnover (PT), which refer to points scored after a turnover made by the opposite team; (d) second chance points (SCP), which refer to points scored after an offensive rebound; (e) points from the bench (PB), which refer to the amount of points scored by bench players.

2.3. Statistical Analysis

Games were classified based on their score difference through a hierarchical cluster analysis using Ward's method and the Squared Euclidian distance as interval. The game classification through cluster analysis has been previously used in literature since it can provide more details on the relevance of the analyzed basketball games [16,17]. The hierarchical cluster analysis was performed using the software SPSS (Version 25.0). A magnitude-based statistics approach was applied to assess the chance of true differences (i.e., greater than the smallest worthwhile change) between winning and losing teams in each cluster for each performance indicator. All data were log-transformed for analysis to reduce bias arising from non-uniformity error and then analyzed for practical significance using magnitude-based inferences on a modified statistical spreadsheet [18]. Data were expressed as mean $\pm$ standard deviation, with pairwise comparisons determined using percentage of mean difference and effect size statistics (Cohen's d) with 90% confidence intervals. The smallest worthwhile change was calculated as a standardized small effect size (0.2) multiplied by the between-subject standard deviation. Chances of real differences in variables were assessed qualitatively as: <1% = almost certainly not; 1–5% = very unlikely; 5–25% = unlikely; 25–75% = possibly; 75–95% = likely; 95–99% = very likely; and >99% = most likely. Clear effects greater than 75% were considered substantial [19]. If the chances of a variable having higher and lower differences were both >5%, the true effect was deemed to be unclear. Effect sizes were rated as follows: <0.20 = trivial; 0.20–0.59 = small; 0.60–1.19 = moderate; 1.20–1.99 = large; and >2.00 = very large [19].

3. Results

Cluster analysis grouped the analyzed games in 18 close, 13 balanced and 9 unbalanced games (score difference: 1–9 points; 10–19 points; 20–33 points, respectively) (Figure 1). The differences between winning and losing teams in each cluster for each performance indicator are shown in Table 1. In close games, winning teams showed a substantially higher number of points in the paints (likely negative) and points from turnover (likely negative), and a lower number of points from the bench (likely positive) compared to losing teams. No substantial differences (unclear) were shown for the other analyzed performance indicators. In balanced games, the only substantial difference found was for second chance points (likely negative). Considering unbalanced games, winning teams revealed a higher number of points in the paint (most likely negative), points from turnover (very likely negative), second chance points (very likely negative), and points from the bench (most likely negative) compared to losing teams.

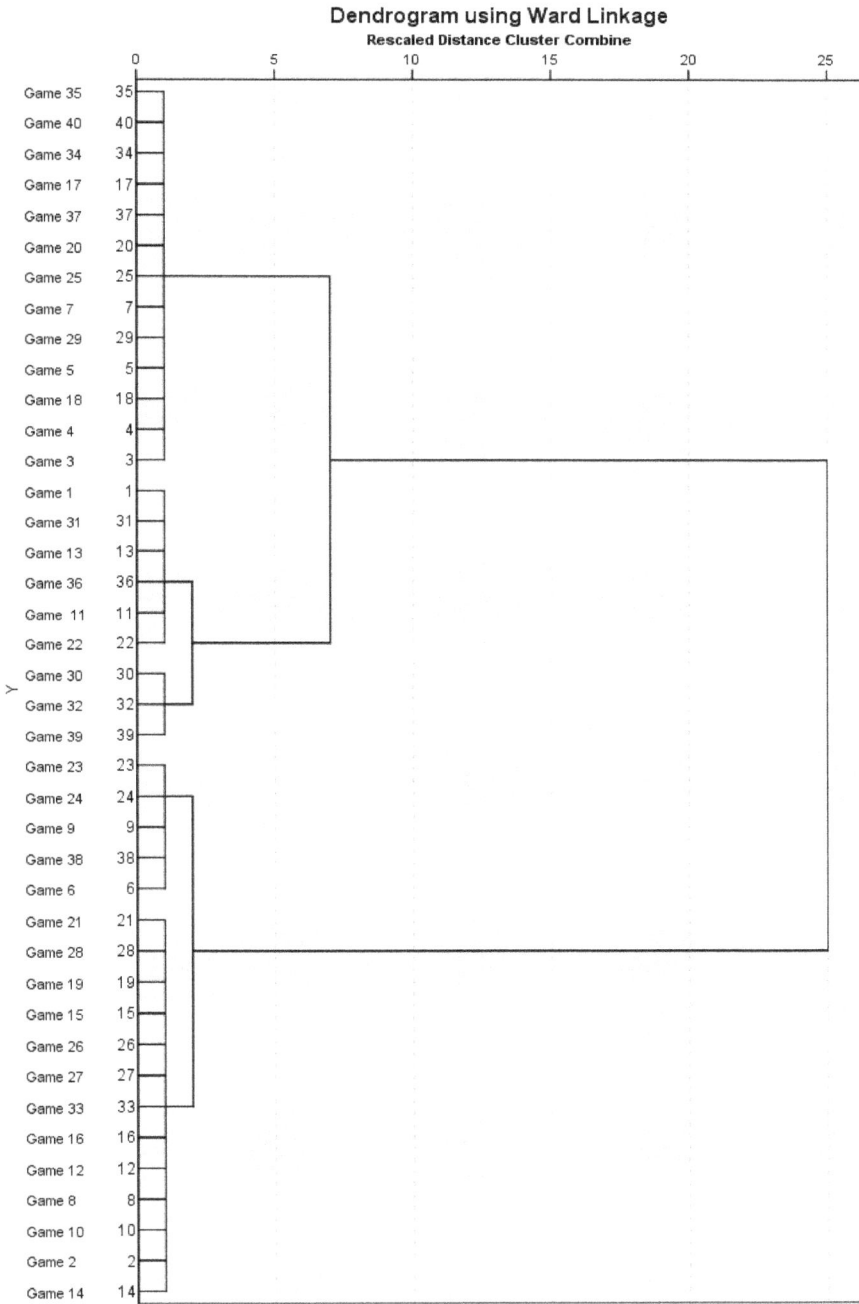

Figure 1. Dendrogram representing the three groups resulting from the hierarchical cluster analysis.

Table 1. Scoring strategies for winning and losing teams in relation to different game scores (close, balanced, and unbalanced games) expressed as mean ± standard deviation (SD), percentage (%), mean difference, and effect size (ES) with their 90% confidence intervals (CI) and magnitude-based inference.

Clusters	Scoring Strategies	Game Outcome		Losing vs. Winning Teams Comparisons		
		Winning Teams	Losing Teams	Mean Difference (90% CI)	ES (90% CI)	Magnitude-Based Inference
Close games	Fast break points	7.6 ± 4.4	6.8 ± 3.2	−0.7 (−2.9; 1.5)	−0.17 (−0.73; 0.39)	Unclear (14/40/47)
	Points in the paint	30.1 ± 6.2	26.3 ± 6.4	−3.8 (−7.3; −0.2)	−0.55 (−1.10; 0.00)	Likely negative (1/13/85)
	Points from turnover	12.8 ± 4.1	10.8 ± 3.7	−1.9 (−4.1; 0.3)	−0.43 (−0.98; 0.12)	Likely negative (3/21/76)
	Second chance points	7.2 ± 3.9	7.2 ± 3.6	−0.1 (−2.2; 2.0)	−0.10 (−0.66; 0.46)	Unclear (19/43/38)
	Points from the bench	13.1 ± 6.8	18.1 ± 9.9	5.1 (0.3; 9.8)	0.56 (0.01; 1.11)	Likely positive (86/12/1)
Balanced games	Fast break points	7.8 ± 3.0	6.5 ± 4.2	−1.2 (−3.7; 1.2)	−0.27 (−0.94; 0.40)	Unclear (12/31/57)
	Points in the paint	28.3 ± 7.7	27.1 ± 6.4	−1.2 (−6.0; 3.5)	−0.13 (−0.78; 0.52)	Unclear (19/37/43)
	Points from turnover	13.2 ± 5.3	12.5 ± 7.1	−0.8 (−5.0; 3.5)	−0.24 (−0.89; 0.41)	Unclear (13/33/54)
	Second chance points	10.0 ± 4.9	7.0 ± 2.3	−3.0 (−5.6; −0.4)	−0.53 (−1.18; 0.12)	Likely negative (3/16/80)
	Points from the bench	18.5 ± 9.1	21.6 ± 9.4	3.1 (−3.1; 9.3)	0.38 (−0.27; 1.04)	Unclear (68/25/7)
Unbalanced games	Fast break points	7.1 ± 5.3	5.3 ± 5.1	−1.8 (−6.1; 2.5)	−0.23 (−1.04; 0.58)	Unclear (18/29/53)
	Points in the paint	30.4 ± 5.5	19.1 ± 4.1	−11.3 (−15.4; −7.3)	−2.18 (−2.97; −1.39)	Most likely negative (0/0/100)
	Points from turnover	15.1 ± 4.8	8.6 ± 6.4	−6.6 (−11.3; −1.8)	−1.04 (−1.87; −0.21)	Very likely negative (1/4/95)
	Second chance points	9.3 ± 4.5	5.7 ± 2.0	−3.7 (−6.6; −0.7)	−1.06 (−1.85; −0.27)	Very likely negative (1/3/96)
	Points from the bench	30.3 ± 11.4	12.3 ± 8.9	−18.0 (−26.4; −9.6)	−1.58 (−2.36; −0.79)	Most likely negative (0/0/100)

4. Discussion

The aim of the study was to examine the scoring strategies differentiating between winning and losing teams during FIBA EuroBasket Women 2017 according to final score differences (close, balanced, and unbalanced games). Results revealed that (a) no substantial differences were shown in FBP in any investigated cluster; (b) winning teams showed a substantially higher number of PP and PT (in close and unbalanced games) and SCP (in balanced and unbalanced games) compared to losing teams; (c) winning teams scored substantially lower and higher number of BPs in close games and unbalanced games, respectively, compared to losing teams.

While the game-related statistics differentiating between winning and losing teams have been widely investigated in women's basketball [5–7,12], little information is available on the scoring strategies adopted by these teams during games. Interestingly, an unclear difference was shown in FBPs scored between winning and losing teams. Previous studies demonstrated that the fast break is one of the most important offensive actions differentiating between winning and losing teams in elite men and women's basketball [8,9,11,20]. Indeed, the fast break action is characterized by a high scoring percentage (i.e., 63–73%) since defense is usually outnumbered and/or not properly organized [11,21]. The unclear difference found between winning and losing teams in the FBPs scored indicates that fast break action is not one of the parameters differentiating between winning and losing teams in the EuroBasket Women 2017 championship. This finding might be explained by the tactical strategies adopted during EuroBasket Women 2017. Possibly, both winning and losing teams were performing fewer fast break and more set-offense actions in a tournament scenario like EuroBasket Women 2017, which is characterized by a congested match schedule compared to the national championship [22]. A previous investigation analyzing the tactical demand of tournament and seasonal games demonstrated 16% fewer fast break actions during tournament games and a longer mean duration of ball possessions [22]. The authors of this study suggested that this difference might be attributable to a higher level of the opponents with more developed defensive systems able to deny early scoring opportunities in international tournaments. Moreover, the fast break action requires a high level of physical fitness [11], while in a tournament scenario with a congested match schedule, players might have to slow down their pace to prevent possible fatigue toward the end of the competition [23]. Future investigations should assess whether the fast break action is a parameter discriminating between winning and losing teams in both women's elite tournament and seasonal championships.

The results of our study also identified PP as one of the main indicators differentiating between winning and losing teams particularly in unbalanced and close games. This result might be explained by the possible importance of inside games in women's basketball. The interaction between outside and inside players has been suggested to be a crucial element in European basketball and in NBA [10,24]. Indeed, Courel et al. [10] demonstrated an increase in the effectiveness of ball possessions including the inside pass from 49.8% to 63.3% in the Spanish professional male league. The inside game has been suggested to be fundamental in discriminating between winning and losing teams also in college basketball due to a substantially higher number of post entries (i.e., a pass from another position to the post area) documented by winning teams [25]. The importance of the inside games has been also documented in women's basketball [26]. Gomez et al. [26] showed that the action completed in the key area reported the highest effectiveness in the women's professional basketball league. Therefore, the results of our investigation possibly substantiate the importance of playing the inside game tactics in elite women's basketball.

A further scoring strategy adopted substantially more by winning teams regards the PT. This result might be a consequence of the fact that losing teams performed more turnovers during the games. Indeed previous investigations analyzing the game-related statistics highlighted that turnover is the main parameter differentiating between winning and losing teams in women's basketball [7,12]. Thus, our result confirms this idea that turnover possibly creates many scoring opportunities for the opponents teams.

The analysis of SCP demonstrated that although winning teams scored a substantially higher number of points deriving from second chances in unbalanced and balanced games, an unclear difference was shown in close games. Previously, Gomez et al. [26] highlighted that elite women's basketball teams obtained a higher offensive effectiveness when starting their attack in the offensive key area, probably due to offensive rebounds [26]. Conversely, a previous investigation analyzing the number of offensive rebounds in winning and losing college teams in close games documented an unclear difference [25]. Therefore, our findings possibly substantiate this result, highlighting that points scored from a second chance (i.e., mainly offensive rebounds) might be not a discriminant parameter between winning and losing teams in basketball close games. Considering these results, further studies should investigate this issue.

The analysis of PB highlighted contrasting results in unbalanced and close games. Winning teams scored a substantially higher number of PB compared to losing teams in unbalanced games, possibly due to the use of more bench players for winning teams during the garbage time likely to allow their best players to recover for the subsequent phases of the tournament. Conversely, winning teams showed a substantially lower number of points scored by bench players compared to losing teams in close games. A possible explanation for our result may be that losing teams were substituting more players, possibly to recover from the disadvantaged situation. This would allow more playing time for bench players, allowing them to score more. Indeed, playing time has been shown to be positively related to shooting performance in the male 1st division Spanish championship [14]. Therefore, our results possibly substantiate the importance of high-quality bench players, and call for future studies investigating their scoring effectiveness in relation to playing time in women's basketball.

Although this study provides new information regarding the scoring strategies differentiating between winning and losing teams, it presents some limitations. Indeed, it only focused on the points scored but not effectiveness of the investigated actions. Moreover, the use of different statistical procedures might provide new insights regarding the association between the investigated scoring strategies and the possibility to win. Therefore, future studies should focus on investigating the effectiveness of the fast break, inside game, actions deriving from turnovers and offensive rebounds, and of substituting players using the notational analysis technique and further statistical approaches such as binary logistic regression or the conditional inference classification tree.

In conclusion, this study provides information on some of the most adopted scoring strategies differentiating between elite women's winning and losing teams according to different game scores. Overall, FBP do not differentiate between winning and losing teams in each investigated cluster. All the other investigated scoring strategies differentiate between winning and losing teams in unbalanced and close games, except for SCP, which demonstrated an unclear difference in close games. These findings might provide useful information for basketball coaches to optimize their training sessions and game strategies.

Author Contributions: Data curation, I.L.; Methodology, D.C. and I.L.; Supervision, D.C.; Writing–original draft, D.C. and I.L.; Writing–review & editing, D.C.

Funding: This research received no external funding

Conflicts of Interest: The authors declare no conflict of interest

References

1. Chen, S.; Duncan, T.; Street, E.; Hesterberg, B. Differences in Official Athletic Website Coverage and Social Media Use between Men's and Women's Basketball Teams. *Sport J.* **2016**, *49*. Available online: http://thesportjournal.org/article/differences-in-official-athleticwebsite-coverage-and-social-media-use-between-mens-and-womens-basketball-teams/ (accessed on 28 April 2018).
2. Conte, D.; Favero, T.G.; Lupo, C.; Francioni, F.M.; Capranica, L.; Tessitore, A. Time-motion analysis of Italian elite women's basketball games: Individual and team analyses. *J. Strength Cond. Res.* **2015**, *29*, 144–150. [CrossRef] [PubMed]

3. Scanlan, A.T.; Dascombe, B.J.; Reaburn, P.; Dalbo, V.J. The physiological and activity demands experienced by Australian female basketball players during competition. *J. Sci. Med. Sport* **2012**, *15*, 341–347. [CrossRef] [PubMed]
4. Bazanov, B.; Rannama, I. Analysis of the offensive teamwork intensity in elite female basketball. *JHSE* **2015**, *10*, 47–51. [CrossRef]
5. Dimitros, E.; Garopoulou, V.; Bakirtzoglou, P.; Maltezos, C. Differences and discriminant analysis by location in A1 Greek women's basketball league. *Sport Sci.* **2013**, *6*, 33–37.
6. Gómez, M.; Lorenzo, A.; Sampaio, J.; Ibáñez, S. Differences in game-related statistics between winning and losing teams in women's basketball. *J. Hum. Mov. Stud.* **2006**, *51*, 357–369.
7. Leicht, A.S.; Gomez, M.A.; Woods, C.T. Team Performance Indicators Explain Outcome during Women's Basketball Matches at the Olympic Games. *Sports* **2017**, *5*, 96. [CrossRef]
8. Román, I.R.; Durán, I.U.R.; Molinuevo, J.S. Analysis of men's and women's basketball fast-breaks. *La RPD* **2009**, *18*, 439–444.
9. Ortega, E.; Palao, J.M.; Gómez, M.Á.; Lorenzo, A.; Cárdenas, D. Analysis of the efficacy of possessions in boys' 16-and-under basketball teams: Differences between winning and losing teams. *Percept. Mot. Skills* **2007**, *104*, 961–964. [CrossRef] [PubMed]
10. Courel-Ibáñez, J.; McRobert, A.P.; Toro, E.O.; Vélez, D.C. Inside pass predicts ball possession effectiveness in NBA basketball. *Int. J. Perform. Anal. Sport* **2016**, *16*, 711–725. [CrossRef]
11. Conte, D.; Favero, T.; Niederhausen, M.; Capranica, L.; Tessitore, A. Determinants of the effectiveness of fast break actions in elite and sub-elite Italian men's basketball games. *Biol. Sport* **2017**, *34*, 177–183. [CrossRef] [PubMed]
12. Koon Teck, K.; Wang, C.; Mallett, C. Discriminating factors between successful and unsuccessful elite youth Olympic Female basketball teams. *Int. J. Perform. Anal. Sport* **2012**, *12*, 119–131. [CrossRef]
13. Sampaio, J.; Ibáñez, S.; Lorenzo, A.; Gómez, M. Discriminative game-related statistics between basketball starters and nonstarters when related to team quality and game outcome. *Percept. Mot. Skills* **2006**, *103*, 486–494. [CrossRef] [PubMed]
14. Sampaio, J.; Drinkwater, E.J.; Leite, N.M. Effects of season period, team quality, and playing time on basketball players' game-related statistics. *Eur. J. Sport Sci.* **2010**, *10*, 141–149. [CrossRef]
15. Harriss, D.; Atkinson, G. Ethical standards in sport and exercise science research: 2014 update. *Int. J. Sports Med.* **2013**, *34*, 1025–1028. [CrossRef] [PubMed]
16. Csataljay, G.; O'Donoghue, P.; Hughes, M.; Dancs, H. Performance indicators that distinguish winning and losing teams in basketball. *Int. J. Perform. Anal. Sport* **2009**, *9*, 60–66. [CrossRef]
17. Lorenzo, A.; Gomez, M.A.; Ortega, E.; Ibanez, S.J.; Sampaio, J. Game related statistics which discriminate between winning and losing under-16 male basketball games. *J. Sports Sci. Med.* **2010**, *9*, 664–668. [PubMed]
18. Hopkins, W. Spreadsheets for analysis of controlled trials with adjustment for a predictor. *Sportscience* **2007**, *11*, 22–23.
19. Hopkins, W.; Marshall, S.; Batterham, A.; Hanin, J. Progressive statistics for studies in sports medicine and exercise science. *Med. Sci. Sports Exerc.* **2009**, *41*, 3–12. [CrossRef] [PubMed]
20. Evangelos, T.; Alexandros, K.; Nikolaos, A. Analysis of fast breaks in basketball. *Int. J. Perform. Anal. Sport* **2005**, *5*, 17–22. [CrossRef]
21. Garefis, A.; Tsitskaris, G.; Mexas, K.; Kyriakou, D. Comparison of the effectiveness of fast breaks in two high level basketball championships. *Int. J. Perform. Anal. Sport* **2007**, *7*, 9–17. [CrossRef]
22. Klusemann, M.J.; Pyne, D.B.; Hopkins, W.G.; Drinkwater, E.J. Activity profiles and demands of seasonal and tournament basketball competition. *Int. J. Sports Physiol.* **2013**, *8*, 623–629. [CrossRef]
23. Montgomery, P.G.; Pyne, D.B.; Hopkins, W.G.; Dorman, J.C.; Cook, K.; Minahan, C.L. The effect of recovery strategies on physical performance and cumulative fatigue in competitive basketball. *J. Sports Sci.* **2008**, *26*, 1135–1145. [CrossRef] [PubMed]
24. Bourbousson, J.; Sève, C.; McGarry, T. Space–time coordination dynamics in basketball: Part 1. Intra-and inter-couplings among player dyads. *J. Sports Sci.* **2010**, *28*, 339–347. [CrossRef] [PubMed]

25. Conte, D.; Tessitore, A.; Gjullin, A.; Mackinnon, D.; Lupo, C.; Favero, T. Investigating the game-related statistics and tactical profile in NCAA division I men's basketball games. *Biol. Sport* **2018**, *35*, 137–143. [CrossRef]

26. Gómez, M.; Lorenzo, A.; Ibañez, S.; Sampaio, J. Ball possession effectiveness in men's and women's elite basketball according to situational variables in different game periods. *J. Sports Sci.* **2013**, *31*, 1578–1587. [CrossRef] [PubMed]

© 2018 by the authors. Licensee MDPI, Basel, Switzerland. This article is an open access article distributed under the terms and conditions of the Creative Commons Attribution (CC BY) license (http://creativecommons.org/licenses/by/4.0/).

sports

MDPI

Article

Seasonal and Longitudinal Changes in Body Composition by Sport-Position in NCAA Division I Basketball Athletes

Jennifer B. Fields [1,2,*], Justin J. Merrigan [1,2], Jason B. White [1,2] and Margaret T. Jones [1,2] (ORCID)

[1]   Frank Pettrone Center for Sports Performance, George Mason University, Fairfax, VA 22030, USA; jmerrig2@gmu.edu (J.J.M.); jwhite35@gmu.edu (J.B.W.); mjones15@gmu.edu (M.T.J.)
[2]   Division of Health and Human Performance, George Mason University, Manassas, VA 20110, USA
*   Correspondence: Jfields8@gmu.edu

Received: 23 July 2018; Accepted: 20 August 2018; Published: 22 August 2018

Abstract: The purpose of this study was to assess the body composition of male and female basketball athletes ($n = 323$) across season, year, and sport-position using air displacement plethysmography. An independent sample t-test assessed sport-position differences. An analysis of variance was used to assess within-subjects across season (pre-season, in-season, and off-season), and academic year (freshman, sophomore, and junior). For both men and women basketball (MBB, WBB) athletes, guards had the lowest body fat, fat mass, fat free mass, and body mass. No seasonal differences were observed in MBB, but following in-season play for WBB, a reduction of ($p = 0.03$) in fat free mass (FFM) was observed. Across years, MBB showed an increase in FFM from freshman to sophomore year, yet remained unchanged through junior year. For WBB across years, no differences occurred for body mass (BM), body fat (BF%), and fat mass (FM), yet FFM increased from sophomore to junior year ($p = 0.009$). Sport-position differences exist in MBB and WBB: Guards were found to be smaller and leaner than forwards. Due to the importance of body composition (BC) on athletic performance, along with seasonal and longitudinal shifts in BC, strength and conditioning practitioners should periodically assess athletes BC to ensure preservation of FFM. Training and nutrition programming can then be adjusted in response to changes in BC.

Keywords: body fat; collegiate athletes; fat free mass; women athletes

1. Introduction

Body composition (BC) plays a critical role in athlete health and sport performance. Extreme levels of body fat (BF%) may bring about severe health consequences. Low BF% has been related to decreased bone density, menstrual dysfunction, and disordered eating habits; high BF% has been related to the onset of cardiovascular disease risk factors. Generally, lesser amounts of fat mass (FM) coupled with greater amounts of fat free mass (FFM), particularly muscle mass, are favorable for athletes [1] and provide the basic foundation for sport-specific technical skills and locomotor activities [2]. The specific balance of FM and FFM, or overall BF%, may be dependent upon sport-position. For example, when evaluating BC across positions in collegiate and elite level basketball players [3–5], guards were reported to be smaller-bodied with lower BF% and FM when compared to centers and forwards. Yet, prior studies have lacked sufficient sample sizes, which allowed for enhanced generalizability and a better understanding of sport-position BC measures, as well as possible performance evaluations.

In addition, routine monitoring of BC in athletes is important to evaluate health assessments, track changes, and make necessary adjustments to a diet or training program [6]. Despite the health and performance implications of BC, few studies have examined both seasonal and longitudinal BC changes in National Collegiate Athletic Association Division I (NCAA DI) men and women basketball (MBB, WBB) athletes.

Consistent and effective monitoring of BC across basketball seasons and years may provide coaches with beneficial feedback in regard to evaluation of strength and conditioning programs and athlete diets. Ultimately, this will enable coaches to make specific adjustments to achieve and maintain optimal BC measures. Prior investigations examining seasonal BC changes in MBB athletes are minimal [7,8], and the few published studies provide contrasting results. Groves et al. (1993) reported a reduction in BF% and BM from pre-season to off-season for NCAA MBB athletes ($n = 8$), while Hoffman et al. (1991) found no change in BF% or BM from pre-season to off-season in MBB athletes ($n = 9$). Both studies had limited sample sizes and used skinfolds to predict BF% [7,8]. While also limited, seasonal BC reports in WBB have shown a reduction in BF% from pre- to post-season [1,9,10]. These limited results warrant the need for further research relative to basketball athletes regarding seasonal BC change.

Furthermore, a unique component of our study was the longitudinal assessment of BC. Few studies have examined BC changes across years in collegiate MBB and WBB athletes [1,11], highlighting the need for additional research that investigates these BC changes. Therefore, the primary purpose of the current study was (1) to provide descriptive data across sport and sport-position in NCAA-DI MBB and WBB athletes; (2) to examine seasonal changes in BC measures; and (3) to document yearly changes in BC measures. We hypothesized that guards would be smaller and leaner compared to forwards, and that MBB and WBB would show reductions in BF% from pre- to post-season, while gaining FFM across years.

2. Materials and Methods

2.1. Subjects

NCAA DI men and women basketball players, aged 18–24, participated in the study. All athletes were under the direction of a strength and conditioning coach and were following sport-specific training regimens with neuromuscular demands particular to their respective sport and training program. Furthermore, nutritional programming was provided by George Mason University's basketball sports dietitian. All participants completed a medical history form and had been cleared previously for intercollegiate athletic participation. Risks and benefits were explained to athletes and an institutionally approved consent form was signed prior to participation. The Institutional Review Board for Human Subjects approved all procedures.

2.2. Procedures

In order to obtain sport-position specific BC data, body composition was assessed over an eight-year period for MBB and a nine-year period for WBB (MBB, $n = 127$; WBB, $n = 196$) using air displacement plethysmography at one-time point per athlete. The MBB and WBB athletes (MBB, $n = 16$; WBB, $n = 29$) that completed three BC assessments within the same year (pre-season: November; in-season: March; off-season: July) were included in a secondary analysis to assess seasonal changes. Furthermore, those who completed BC assessments (MBB, $n = 14$; WBB, $n = 8$) across three consecutive years (freshman, sophomore, and junior) were included in an analysis of longitudinal BC changes. Longitudinal measurements were separated by a 12-month period. Differences in BF%, FM, FFM, and BM were evaluated. Findings were compared across sport-position, seasons, and years.

Athletes were instructed to refrain from exercise, eating, and drinking for ≥ 2 h prior to testing. The majority of testing was conducted early in the morning following an overnight fast. Upon arrival to the laboratory, subjects' body mass was recorded to the nearest 0.01 cm and 0.02 kg, respectively, using a stadiometer (Detecto, Webb City, MO, USA) and electronic scale (BOD POD; COSMED USA Inc., Concord, CA, USA) calibrated according to manufacturer guidelines. Body composition was assessed using air displacement plethysmography (BOD POD, model 2000A; COSMED USA Inc.,

Concord, CA, USA), which has been shown to be a reliable and valid method for measuring BC [12]. Prior to each testing session, calibration procedures were completed using an empty chamber and a calibrating cylinder of a standard volume (49.55 L), according to the manufacturer guidelines. Participants were instructed to wear a formfitting sports bra (women), spandex shorts, and swim cap, and to remove all jewelry, in accordance with standard operating procedures that reduced excess air displacement. A trained technician performed all testing. Two tests were performed to ensure reliability of the assessment. If the tests results were not within 150 mL of each other, two additional tests were administered. Our lab's test to test reliability of this body composition assessment yielded high reliability for body mass (r = 1.0), body fat percent (r = 0.997), and fat-free mass (r = 1.0) [13].

2.3. Statistical Analysis

SPSS version 25.0 (IBM, Armonk, NY, USA) was used for data analysis. To test for significant differences across sport-position, we used an independent samples t-test. To assess seasonal (pre-season, in-season, and off-season) and longitudinal (freshman, sophomore, and junior) changes, we used a repeated measures ANOVA.

3. Results

Body compositions between MBB and WBB sport-positions are included in Table 1 (means ± SD; BF%, FM, FFM, and BM). For both MBB and WBB, guards, when compared to forwards, had significantly ($p < 0.001$) less BF% (MBB: 8.6 ± 3.3% vs. 14.9 ± 4.8%; WBB: 19.2 ± 6.3% vs. 24.2 ± 5.7%), FM (MBB: 7.4 ± 3.1 kg vs. 15.9 ± 5.7 kg; WBB: 13.4 ± 5.4 kg vs. 20.5 ± 7.4 kg), FFM (MBB: 77.7 ± 6.4 kg vs. 89.4 ± 7.5 kg; WBB: 54.63 ± 64.4 kg vs. 61.8 ± 6.0 kg), and BM (MBB: 85.2 ± 7.5 kg vs. 105.3 ± 8.1 kg; WBB: 68.0 ± 7.4 kg vs. 82.2 ± 12.5 kg). After adjusting for height, however, FFM differences were no longer apparent (MBB: 22.2 ± 1.8 vs. 22.0 ± 1.4; WBB: 18.6 ± 1.7 vs. 18.6 ± 1.9).

Table 1. Body composition between basketball sport-positions.

Sex	Position	BF (%)	FM (kg)	FFM (kg)	BM (kg)	Height (cm)	FMI (kg/m^2)	FFMI (kg/m^2)
Men	Guard (n = 68)	8.6 ± 3.3 [1]	7.4 ± 3.1 [1]	77.7 ± 6.4 [1]	85.2 ± 7.4 [1]	187.4 ± 7.0	2.1 ± 0.9 [1]	22.2 ± 1.8
	Forwards (n = 59)	14.9 ± 4.8 [2]	15.9 ± 5.6 [2]	89.5 ± 5.9 [2]	105.3 ± 8.0 [2]	201.7 ± 4.0	3.9 ± 1.4 [2]	22.0 ± 1.4
Women	Guard (n = 105)	19.2 ± 6.3 [1]	13.4 ± 5.4 [1]	54.6 ± 4.4 [1]	68.0 ± 7.4 [1]	171.6 ± 5.0 [1]	4.52 ± 1.8 [1]	18.6 ± 1.7
	Forwards (n = 91)	24.2 ± 5.7 [2]	20.5 ± 7.7 [2]	61.8 ± 5.9 [2]	82.2 ± 12.5 [2]	183.5 ± 4.4 [2]	6.07 ± 2.3 [2]	18.6 ± 1.9

Values are mean ± SD; BF%: Body fat percent; FM: Fat mass; FFM: Fat free mass; BM: Body mass; FMI: Fat mass index; FFMI: Fat free mass index; Order of significance presented: [1] < [2]; Level of significance set at $p < 0.0125$.

For season phases, there were no significant differences in BF%, FM, FFM, or BM observed in MBB (Table 2). Following in-season play, a significant reduction in FFM ($p = 0.01$) was observed for WBB (March: 58.8 ± 6.5 kg vs. July: 56.9 ± 6.7 kg) (mean difference; 95% CI: −0.814; −1.601, −0.027) (Table 2).

When analyzing MBB body composition changes across years, FFM increased significantly ($p < 0.05$) from freshman (82.0 ± 8.4 kg) to sophomore year (83.5 ± 8.4 kg) (mean difference; 95% CI: 1.534; 0.008, 3.059) and remained unchanged through junior year (84.1 ± 9.1 kg) (mean difference; 95% CI: 0.618; −0.352, 1.587) (Table 3). For WBB, no differences were observed across years in BM, BF%, and FM. FFM did not increase from freshman to sophomore years (mean difference; 95% CI: 0.918; −0.891, 2.728), but significant ($p = 0.02$) increases occurred from sophomore (56.2 ± 3.9 kg) to junior year (57.7 ± 4.0 kg) (mean difference; 95% CI: 1.448; 0.259, 2.638) (Table 3).

Table 2. Body composition measures across basketball season phases.

Sex	Measure	Pre-Season	In-Season	Off-Season
Men (*n* = 16)	BF (%)	10.4 ± 5.2	10.0 ± 4.5	10.7 ± 6.2
	FM (kg)	10.3 ± 6.3	9.7 ± 5.5	10.7 ± 7.5
	FFM (kg)	83.4 ± 9.2	83.6 ± 8.8	83.1 ± 9.2
	BM (kg)	93.7 ± 14.1	93.3 ± 13.0	93.3 ± 14.5
Women (*n* = 29)	BF (%)	16.8 ± 7.4	17.0 ± 7.7	17.7 ± 8.5
	FM (kg)	21.7 ± 6.4	21.6 ± 6.7	22.2 ± 6.7
	FFM (kg)	57.9 ± 7.7 [1]	58.8 ± 6.5 [2]	56.9 ± 6.6 [1]
	BM (kg)	75.4 ± 13.1	75.9 ± 12.8	75.3 ± 13.5

Values are mean ± SD; BF%: Body fat percent; FM: Fat mass; FFM: Fat free mass; BM: Body mass; Order of significance presented: [1] < [2]; Level of significance set at $p < 0.05$.

Table 3. Body composition measures across years in basketball athletes.

Sex	Measure	Year 1	Year 2	Year 3
Men (*n* = 14)	BF (%)	14.1 ± 4.2	12.6 ± 4.7	13.4 ± 4.5
	FM (kg)	14.0 ± 5.6	12.5 ± 6.1	13.3 ± 5.8
	FFM (kg)	82.0 ± 8.4 [1]	83.5 ± 8.4 [2]	84.1 ± 9.1 [2]
	BM (kg)	95.9 ± 12.8	96.0 ± 12.5	97.4 ± 12.9
Women (*n* = 8)	BF (%)	21.0 ± 5.5	21.1 ± 5.1	20.5 ± 5.3
	FM (kg)	14.8 ± 4.5	15.3 ± 4.7	15.1 ± 4.9
	FFM (kg)	55.2 ± 3.4 [1]	56.1 ± 3.9 [1]	57.7 ± 4.0 [2]
	BM (kg)	70.1 ± 4.6	71.4 ± 6.5	72.8 ± 6.5

Values are mean ± SD; BF%: Body fat percent; FM: Fat mass; FFM: Fat free mass; BM: Body mass; Order of significance presented: [1] < [2]; Level of significance set at $p < 0.05$.

4. Discussion

Previously published body composition data pertaining to sport-position, sport season, and time are limited for NCAA DI men and women basketball athletes. Therefore, our study's purpose was to contribute descriptive BC data for men and women collegiate basketball athletes, and to evaluate seasonal and longitudinal changes in BC metrics. Researchers hypothesized that (1) guards would be smaller and leaner compared to forwards; and (2) MBB and WBB would show reductions in BF% from pre- to post-season, and gain FFM across years.

In the current study, MBB and WBB guards were significantly smaller and leaner than forwards, which is in support of previously published research [3–5]. In basketball, a player's size largely determines the position played on the team [5]. For example, it is advantageous to assign the largest players positions closer to the basket, which enhances shot blocking and rebounding performances. Smaller players, however, are assigned perimeter positions that facilitate moving the ball quickly down the court [14]. The smaller physique is suitable for guards, as the position requires speed and agility skills, an ability to rapidly transfer the ball from defense to offense, and an ability to defend against the quickest players on the opposing team [5,14].

The investigation of BC changes across sport seasons in collegiate basketball athletes is not extensive. In the current study, MBB demonstrated no significant changes in BF%, FM, FFM, or BM from the initiation of pre-season to completion of the in-season. These findings are similar to previous findings in a smaller sample of NCAA D1 MBB athletes (*n* = 9) [8]. However, results appear inconsistent, as Groves and Gayle observed a significant reduction in BF% from the start of pre-season (October) to conclusion of the regular season (April) [7]. Similarly, elite junior MBB athletes have shown a 3.7% and 2.5% increase in FFM and BM over sport seasons, with no change in absolute FM [15]. However, the small sample sizes (*n* = 5) makes generalizing results difficult [15]. Furthermore, each of these studies utilized different BC measurement assessments (i.e., skinfolds and doubly labeled water). Therefore, we recommend that caution be exercised when comparing results from different measurement techniques [16].

In the current study, WBB athletes showed a decrease in FFM following the in-season period, while no changes were observed in BF%, FM, or BM. These results are in contrast with previously reported findings, where WBB athletes ($n = 38$) displayed decreases in BF% from the pre- to post-season, ranging from -0.8% to -1.4% [1,9,10], and increases in FFM and BM in elite junior WBB players ($n = 9$) [15]. The reduction in FFM observed in the current study is of concern, as previous literature has shown a correlation between FFM and bone mineral density [17], strength [18,19], speed [20], and power [18,21]. Reductions in FFM across seasons were also observed in NCAA Division I collegiate softball athletes [17], thus signifying an ongoing concern for women athletes in regard to maintaining FFM.

When analyzing longitudinal data, MBB athletes displayed an increase in FFM from years one to two with no change from years two to three. No other changes in BC were observed. Although a lack of published findings makes comparisons challenging, prior research in collegiate MBB players ($n = 16$) showed an increase in FFM and BM from years one to two ($+2.5 \pm 3.8$ kg, $p < 0.05$) ($n = 19$) [22], and years three to four ($+2.8 \pm 3.4$ kg, $p < 0.01$) [21]. BM also increased in a similar pattern, from years one to two ($+2.4 \pm 3.0$ kg, $p < 0.01$), and years three to four ($+3.7 \pm 2.9$ kg, $p < 0.01$) [22]. The largest change observed between years one and two is likely due to athletes following a structured strength training regimen that they later became adapted to [22,23].

Although data on basketball is limited, longitudinal BC responses have been widely examined in NCAA football athletes [24–26]. Division I wide receivers and defensive backs have shown steady increases in BM, with the largest increases observed from years one to two (82.65 ± 3.42 kg to 87.42 ± 2.75 kg; 7.7% increase) [25,26], and no reported changes in BF%. Linemen have also shown strong increases in BM from years one to two (1.9% gain) [24–26], as well as increases in lean body mass (LBM), growing from 97.89 kg to 101.09 kg in years one to two, and 101.09 kg to 104.55 kg in years two to three [26]. The different BC responses found between skill positions (receivers and defensive backs) and linemen may be due to the athletic nature of each position. Skilled players have been shown to have the lowest BF%, as these positions require strong speed and agility skills. On the other hand, linemen require greater amounts of strength and rely on higher levels of FFM, thus explaining why increases in such areas were noted over the years. The same conclusions can be speculated about in basketball, where the type of playing strategies made by the team may dictate the likelihood of BC changes across seasons. Teams that play a faster style of game without a true center position may not experience significant changes in body composition, a finding similar to what was reported in football skill positions.

In the current study, WBB athletes showed no differences across years for BM, BF%, and FM. FFM did not increase from freshman to sophomore year, but a significant increase occurred from sophomore (56.2 ± 3.9 kg) to junior year (57.7 ± 4.0 kg). Few studies have measured longitudinal BC changes; therefore, more research is needed to understand these growth trends over time. Stanforth et al. (2014) observed an increase in BF% (year 1: $25.8 \pm 0.8\%$; year 2: $25.9 \pm 0.8\%$; year 3: $27.5 \pm 0.9\%$) and FM (year 1: 20.0 ± 0.7 kg; year 2: 20.1 ± 0.8 kg; year 3: 21.8 ± 0.8 kg) across three consecutive years in 38 NCAA Division I WBB athletes [1]. Petko and Hunter, however, saw no changes in BF% from freshman to senior year ($n = 11$) (% change: $0.3 \pm 0.8\%$) [11]. Again, this may be due to the various methods used to evaluate BC, training programs, or the style of game play. In the current study, the increases in FFM observed from sophomore to junior year indicate a beneficial adaptation to training, yet it is surprising no changes were observed from freshmen to sophomore year, a period in which new players are commonly exposed to a structured resistance training program. Future research should consider evaluating BC changes across seasons and years with multiple teams utilizing the same measurement tools for analyses.

In WBB athletes, we found the reduction in FFM across seasons and lack of muscle gain across years a cause for concern. This may, in part, be due to a fear of women appearing overly muscular [27], as well as potential training program differences between MBB and WBB teams. These speculations are based upon the self-reported beliefs of MBB coaches that women are physically inferior to men, lack

Sports **2018**, *6*, 85

commitment in the sporting sphere, and are not capable of training with male athletes [28]. Further, high school strength coaches have expressed concern that women are often not challenged to work as hard athletically as men, perhaps because of the perception that women are "dainty" and "ought not to sweat too hard" [29]. If said beliefs are present, regardless of sport level, it is likely that male strength coaches will treat women athletes differently than their male counterparts. These unfounded perceptions are increasingly apparent in high school athletics, where 50% of coaches for men's sports required their athletes to strength train, and only 9% of coaches for women's sports did the same [30]. In fact, researchers have suggested that an even greater focus on force production should be utilized when training women for power [30]. To combat losses in FFM, it is recommended that strength training be made a priority in resistance exercise programming for female athletes [30,31].

In conclusion, strength and conditioning practitioners must ensure training and nutrition programs are aimed at optimizing muscle development to support performance in basketball athletes, particularly in WBB. Routine monitoring of BC measures may assist in the evaluation of such programs and will allow coaches to make adjustments, as needed.

Author Contributions: Conceptualization, J.B.F. and M.T.J.; Methodology, J.B.F., J.J.M., and M.T.J.; Formal Analysis, J.B.F., J.J.M., J.B.W., and M.T.J.; Investigation, J.B.F., J.J.M., and M.T.J.; Data Curation, J.B.F. and J.J.M.; Writing-Original Draft Preparation, J.B.F.; Writing-Review & Editing, J.B.F., J.J.M., J.B.W., and M.T.J.; Supervision, J.B.W. and M.T.J.

Funding: This research received no external funding.

Acknowledgments: The authors would like to thank the athletic trainers, coaches, and student-athletes from intercollegiate athletics for their contribution to this study.

Conflicts of Interest: The authors declare no conflict of interest.

References

1. Stanforth, P.R.; Crim, B.M.; Stanforth, D. Body composition changes among female division I athletes across the competitive season and over a multiyear time frame. *J. Strength Cond. Res.* **2014**, *28*, 300–307. [CrossRef] [PubMed]
2. Turnagöl, H.H. Body composition and bone mineral density of collegiate American football players. *J. Hum. Kinet.* **2016**, *51*, 103–112. [CrossRef] [PubMed]
3. LaMonte, M.J.; McKinnex, J.T.; Quinn, S.M.; Brainbridge, C.N.; Eisenman, P.A. Comparison of physical and physiological variables for female collegiate basketball players. *J. Strength Cond. Res.* **1999**, *13*, 264–270.
4. Latin, R.W.; Berg, T.; Baechle, T. Physical and performance characteristics of NCAA division I male basketball players. *J. Strength Cond. Res.* **1994**, *8*, 214–218. [CrossRef]
5. Ostojic, S.; Masic, S.; Dikic, N. Profiling in basketball. *J. Strength Cond. Res.* **2006**, *20*, 740–744. [CrossRef]
6. Thomas, T.D.; Erdman, K.A.; Burke, L.M. Position of the Academy of Nutrition and Dietetics, Dietitians of Canada, and the American College of Sports Medicine: Nutrition and athletic performance. *J. Acad. Nutr. Diet.* **2016**, *116*, 501–528. [CrossRef] [PubMed]
7. Groves, B.R.; Gayle, R.C. Physiological changes in male basketball players in year-round strength training. *J. Strength Cond. Res.* **1993**, *7*, 30–33.
8. Hoffman, J.R.; Fry, A.C.; Howard, R.; Maresh, C.M.; Kraemer, W.J. Strength, speed and endurance changes during the course of a Division I basketball season. *J. Strength Cond. Res.* **1991**, *5*, 144–149.
9. Carbuhn, A.F.; Fernandex, T.E.; Bragg, A.F.; Green, J.S.; Crouse, S.F. Sport and training influence bone and body composition in women collegiate athletes. *J. Strength Cond. Res.* **2010**, *24*, 1710–1717. [CrossRef] [PubMed]
10. Johnson, G.O.; Nebelsick-Gullett, L.J.; Thorland, W.G.; Housh, T.J. The effect of a competitive season on the body composition of university female athletes. *J. Sports Med. Phys. Fitness* **1989**, *29*, 314–320. [PubMed]
11. Petko, M.A.; Hunter, G.R. Four-year changes in strength, power, and aerobic fitness in women college basketball players. *Strength Cond. J.* **1997**, *19*, 46–49. [CrossRef]
12. Ferri-Morales, A.; Nascimento-Ferreira, V.; Vlachopoulos, D.; Ubago-Guisado, E.; Torres-Costoso, A.; De Moraes, A.C.F.; Gracia-Marco, L. Agreement between standard body composition methods to estimate percentage body fat in young male athletes. *Pediatr. Exerc. Sci.* **2018**. [CrossRef] [PubMed]

13. Fields, J.B.; Metoyer, C.J.; Casey, J.C.; Esco, M.R.; Jagim, A.R.; Jones, M.T. Comparison of body composition variables across a large sample of NCAA women athletes from six competitive sports. *J. Strength Cond. Res.* **2017**. [CrossRef] [PubMed]

14. Drinkwater, E.J.; Pyne, D.B.; McKenna, M.J. Design and interpretation of anthropometric and fitness testing of basketball players. *Sports Med.* **2008**, *38*, 565–578. [CrossRef] [PubMed]

15. Silva, A.M.; Santos, D.A.; Matias, C.N.; Rocha, P.M.; Petroski, E.L.; Minderico, C.S.; Sardinha, L.B. Changes in regional body composition explain increases in energy expenditure in elite junior basketball players over the season. *Eur. J. Appl. Physiol.* **2012**, *112*, 2727–2737. [CrossRef] [PubMed]

16. Vescovi, J.D.; Hildebrandt, L.; Miller, W.; Hammer, R.; Spiller, A. Evaluation of the BOD POD for estimating percent fat in female college athletes. *J. Strength Cond. Res.* **2002**, *16*, 599–605. [CrossRef] [PubMed]

17. Peart, A.; Wadsworth, D.; Washington, J.; Oliver, G. Body composition assessment in National Collegiate Athletic Association Division I softball athletes as a function of playing position across a multiyear time frame. *J. Strength Cond. Res.* **2018**. [CrossRef] [PubMed]

18. Jones, M.T.; Jagim, A.R.; Haff, G.G.; Carr, P.J.; Martin, J.; Oliver, J.M. Greater strength drives difference in power between sexes in the conventional deadlift exercise. *Sports* **2016**, *4*, 43. [CrossRef] [PubMed]

19. Guimarães, B.R.; Pimenta, L.D.; Massini, D.A.; dos Santos, D.; da Cruz Siqueira, L.O.; Simionato, A.R.; dos Santon, L.G.A.; Neiva, C.M.; Filho, D.M.P. Muscle strength and regional lean body mass influence on mineral bone health in young male adults. *PLoS ONE* **2018**, *13*, 1–13. [CrossRef] [PubMed]

20. Brocherie, F.; Girard, O.; Forchino, F.; Millet, G.P. Relationships between anthropometric measures and athletic performance, with special reference to repeated-sprint ability, in the Qatar national soccer team. *J. Sport Sci.* **2014**, *32*, 1–12. [CrossRef] [PubMed]

21. Perroni, F.; Vetrano, M.; Rainoldi, A.; Guidetti, L.; Baldari, C. Relationship among explosive power, body fat, fat free mass and pubertal development in youth soccer players: A preliminary study. *Sports Sci. Health* **2014**, *10*, 67–73. [CrossRef]

22. Hunter, G.R.; Hilyer, J.; Forster, M.A. Changes in fitness during 4 years of intercollegiate basketball. *J. Strength Cond. Res.* **1993**, *7*, 26–29. [CrossRef]

23. Hass, C.J.; Feigenbaum, M.S.; Franklin, B.A. Prescription of resistance training for healthy populations. *Sports Med.* **2001**, *31*, 953–964. [CrossRef] [PubMed]

24. Hoffman, J.R.; Ratamess, N.A.; Kang, J. Performance changes during a college playing career in NCAA division III football athletes. *J. Strength Cond. Res.* **2011**, *25*, 2351–2357. [CrossRef] [PubMed]

25. Jacobson, B.H.; Conchola, E.G.; Glass, R.G.; Thompson, B.J. Longitudinal Morphological and Performance Profiles for American, NCAA Division I Football Players. *J. Strength Cond. Res.* **2013**, *27*, 2347–2354. [CrossRef] [PubMed]

26. Stodden, D.F.; Galitski, H.M. Longitudinal Effects of a Collegiate Strength and Conditioning Program in American Football. *J. Strength Cond. Res.* **2010**, *24*, 2300–2308. [CrossRef] [PubMed]

27. Bennett, E.V.; Scarlett, L.; Clarke, L.H.; Crocker, P.R.E. Negotiating (athletic) femininity: The body and identity in elite female basketball players. *Qual. Res. Sport Exerc.* **2016**, *9*, 233–246. [CrossRef]

28. Tomlinson, A.; Yorganci, I. Male coach/female athlete relations: Gender and power relations in competitive sport. *J. Sport Soc. Issues* **1997**, *21*, 134–155. [CrossRef]

29. Reynolds, M.L.; Ransdell, L.B.; Lucas, S.M.; Petlichkoff, L.M.; Gao, Y. An examination of current practices and gender differences in strength and conditioning in a sample of varsity high school athletic programs. *J. Strength Cond. Res.* **2012**, *26*, 174–183. [CrossRef] [PubMed]

30. Mata, D.J.; Oliver, J.M.; Jagim, A.R.; Jones, M.T. Sex differences in strength and power support the use of a mixed-model approach to resistance training programing. *Strength Cond. J.* **2016**, *38*, 2–7. [CrossRef]

31. Faigenbaum, A. Age- and sex-related differences and their implications for resistance exercise. In *Essentials of Strength Training and Conditioning*, 2nd ed.; Baechle, T.R., Earle, R.W., Eds.; Human Kinetics: Champaign, IL, USA, 2000; pp. 169–186.

© 2018 by the authors. Licensee MDPI, Basel, Switzerland. This article is an open access article distributed under the terms and conditions of the Creative Commons Attribution (CC BY) license (http://creativecommons.org/licenses/by/4.0/).

sports

MDPI

Article

The Negative Influence of Air Travel on Health and Performance in the National Basketball Association: A Narrative Review

Thomas Huyghe [1,*] [iD], Aaron T. Scanlan [2] [iD], Vincent J. Dalbo [2] [iD] and Julio Calleja-González [3]

[1]   BC Oostende, Oostende 8400, Belgium
[2]   Human Exercise and Training Laboratory, School of Health, Medical and Applied Sciences, Central Queensland University, Rockhampton 4702, Australia; a.scanlan@cqu.edu.au (A.T.S.); v.dalbo@cqu.edu.au (V.J.D.)
[3]   Physical Education and Sport Department, University of Basque Country (UPV-EHU), Vitoria-Gasteiz 48940, Spain; Julio.calleja.gonzalez@gmail.com
*   Correspondence: thomashuyghe@hotmail.com; Tel.: +32-470-077-2716

Received: 3 July 2018; Accepted: 24 August 2018; Published: 30 August 2018

Abstract: Air travel requirements are a concern for National Basketball Association (NBA) coaches, players, and owners, as sport-based research has demonstrated short-haul flights ($\leq$6 h) increase injury risk and impede performance. However, examination of the impact of air travel on player health and performance specifically in the NBA is scarce. Therefore, we conducted a narrative review of literature examining the influence of air travel on health and performance in team sport athletes with suggestions for future research directions in the NBA. Prominent empirical findings and practical recommendations are highlighted pertaining to sleep, nutrition, recovery, and scheduling strategies to alleviate the negative effects of air travel on health and performance in NBA players.

Keywords: NBA; athletic performance; fatigue; circadian rhythm; injury; sleep

---

1. National Basketball Association: Schedule and Travel Requirements

The National Basketball Association (NBA) is the premier basketball league in the world [1,2] and in recent years a greater emphasis has been placed on player safety [3,4]. In regard to player safety, there has been increased attention in the areas of training load [3,5] as well as schedule and travel requirements [5]. In an attempt to reduce the training load and schedule requirements of players, the NBA has modified the preseason schedule. Prior to 2017, NBA teams played eight preseason games across 3–4 weeks in preparation for the regular season [6,7]. Since the 2017–2018 season, the NBA season has consisted of four to six preseason games played across 3–4 weeks followed by an 82-game regular season played across 26 weeks (177 days). During the regular season, each team plays two to five games per week (~3.2 games per week) [1] with games lasting an average duration of 2 h and 15 min [2]. NBA teams rarely practice during the season and practices that occur are typically less than 1 h [1,2]. In response to teams resting players during back-to-back (two games within a 2-day span) games [8], the league extended the duration of the regular season by 7 days with the purpose of scheduling fewer back-to-back games [6]. During the 2017–2018 season, NBA teams played an average of 14.4 ± 0.9 back-to-back games, which was the lowest on record compared to any previous season in the NBA [2]. Furthermore, the 2017–2018 NBA season marked the first season in NBA history in which no team played four games in 5 nights [6]. Despite adjustments to the NBA schedule, air travel demands remain high due to the geographical span of teams across four time zones (eastern, central, mountain, and western). In this regard, NBA players spend more time above 30,000 ft than athletes competing in all other team sports in the United States of America (USA) [7]. Air travel requirements are a concern for NBA coaches, players, and owners, as research has demonstrated short-haul flights ($\leq$6 h) increase injury risk [2,9–13] and impede performance [9,14–20]. Competing in away games has been reported to significantly increase regular season injury risk in a sample of 1443 NBA players between 2012 and 2015 [9]. Specifically, 54% of regular season injuries occurred in players playing games away from home, which was significantly greater than the expected injury rate for away games of 50% ($p < 0.05$) [9]. Furthermore, the direction of air travel should be considered by NBA teams, as traveling westward exacerbates reductions in performance [14,21]. In a sample of 8495 NBA games between 1987 and 1995, west coast teams scored four more points per game ($p < 0.05$) when traveling to the east coast than east coast teams scored when traveling to the west coast [21]. Furthermore, NBA teams traveling eastward had a winning percentage of 45.4% compared with 36.2% for teams traveling westward ($p < 0.001$) between 2010 and 2015 [14]. The increased difficulty of traveling westward across the USA to compete has also been reported in the National Football League and the National Hockey League [14]. Westward travel is likely more difficult since performance tends to peak in the late afternoon and players traveling from west to east tend to play games closer to their circadian peaks given most NBA games are played at night.

2. The Impact of Travel Fatigue on Performance

Frequent air travel can negatively affect hydration status, nutritional behaviors, sleep quality, and sleep quantity, thus extending the time for sufficient recovery between games and/or training in athletes [15]. As a result, air travel should be considered as an additional stressor imposed on NBA players in conjunction with competition and training schedules [15], especially when less than 72 h of rest is experienced between games [21,22].

One of the main consequences associated with frequent air travel exposure is "travel fatigue". Travel fatigue refers to feelings of disorientation, light-headedness, gastrointestinal disruption, impatience, lack of energy, and general discomfort that follow traveling across time zones [13]. The magnitude of travel fatigue depends on many factors such as regularity, duration, and conditions of travel [13]. Specific causes of air-related travel fatigue include:

- Prolonged exposure to mild hypoxia [16,23,24].
- Difficulties in standing, walking, and moving around due to limited room inside the air cabin.
- Reduced air quality in the cabin, which may impair immune function [12].
- Dry cabin air and low hypobaric pressure potentially causing dehydration [25].
- Prolonged sitting in a cramped position reducing mobility and flexibility [10,16].
- Disruption of routines (e.g., eating and sleeping) [26].
- Noise of plane and cabin (e.g., sleep disturbance) [16].
- Formalities of air travel may induce negative mood states [26].

A primary issue regarding air travel occurs as a result of significant reductions in oxygen saturation, which has been found to decrease significantly from 97% at ground level to 93% at cruising altitude ($p < 0.05$) [24]. This finding is significant, as oxygen saturation levels of 93% could prompt physicians to administer supplemental oxygen in hospital patients [24] and thus would slow muscle recovery [27]. One study examined the effects of air travel from the east coast to the west coast of the USA on physiological performance measures, sleep quality, and hormonal alterations [28]. However, it is important to note the following: participants used in this investigation were not athletes, a simulated sporting event most closely related to demands experienced during soccer was administered, and there was no non-exercise (control) group. However, air travel induced jet lag symptoms, which resulted in decreased sleep quality and was paired with significantly increased melatonin levels on flight days (travel from east to west coast and travel from west to east coast) [28]. The authors also examined markers of skeletal muscle damage, but since a non-exercise control was not included in the investigation meaningful interpretations of the data cannot be determined [28].

When flying across two or more time zones, symptoms of travel fatigue can remain up to 2–3 days after arrival [13]. The physiological and perceptual stressors associated with flying across one or more time zones may alter sleep patterns in athletes [12]. In particular, short-haul air travel has been reported to impair athletic performance due to the development of an inefficient internally-driven circadian rhythm (i.e., sleep deprivation or disorientation between the circadian system and the environment) [29]. In this sense, NBA players may experience difficulty sleeping at night and excessive daytime sleepiness when traveling across multiple time zones. Subsequently, the greater the number of time zones travelled, the more difficult it is for an athlete to adapt to a new time zone. For example, a 2-h time zone shift may cause marginal disruption to the circadian rhythm, but a 3-h time zone shift (e.g., NBA players traveling coast to coast within the USA) can cause a significant desynchronization of circadian rhythm [13]. Therefore, it is recommended that NBA players focus on physical activity, eating, and social contact during daylight in their new time zone in order to resynchronize their circadian rhythm, especially when traveling from coast to coast [13].

The circadian rhythm plays a critical role in sports performance [13,19,30,31]. When an athlete's circadian rhythm is synchronized with the environment, the athlete should achieve optimal performance during late afternoons and early evenings [19]. Considering air travel can cause an athlete's circadian rhythm to become unsynchronized with the environment, air travel may contribute to the home court advantage in the NBA [32,33], as the body's core temperature (an endogenous measure of circadian rhythm) takes approximately 1 day for each time zone crossed to adapt completely to the new time zone [13,34]. Consequently, the number of time zones traveled plays a critical role in the magnitude of travel fatigue [13].

The regularity, duration, and direction of air travel, combined with in-cabin conditions, likely predisposes NBA players to travel fatigue [13]. In turn, travel fatigue can have deleterious effects on player recovery and subsequent performance, particularly when scheduled soon after practices or games. Consequently, it is recommended that recovery and practices administered before and after air travel are modified to account for travel fatigue, especially considering the travel direction and flight duration experienced.

3. Scheduling and Recovery Opportunities

Besides the direction and duration of air travel, the home court advantage is also influenced by the quantity of rest NBA teams attain prior to games [35]. In particular, a consistent advantage was recorded when a team had more than 1 day of rest between games (the home team's score increased by 1.1 points per game and the away team's score increased by 1.6 points per game) in a sample of 8495 regular season NBA games between 1987–1995 [21]. Moreover, average total scores (home and away teams) were highest when 3 days of rest were encountered between games with data collected from the 1987–1995 seasons [21]. Consequently, the negative influence of air travel during an NBA season may be mitigated by incorporating supplemental days to recover from games.

An optimal recovery window of 72 h following games and practices is needed for an athlete or team to return to optimal levels of performance [22]. Nevertheless, the NBA schedule dictates condensed game schedules that necessitate compressed training schedules, which may inhibit access to active rest days to fully recover from accumulated physical and psychological stress induced by NBA games and practices. Consequently, NBA teams are often obligated to intervene with various ergogenic practices in an attempt to speed up the recovery process, such as whole body cryotherapy, compression tights, cold water immersion, contrast water therapy, and soft tissue massage [36]. While these commonly employed recovery practices, including compression tights [37], cold water immersion [38], and massage [39], have been investigated in various samples of basketball players, no data are available specifically in NBA players. Therefore, more research is needed to ascertain if these recovery practices benefit NBA players across the season.

Another factor to consider in reducing injury risk and optimizing performance in the NBA is the total amount of in-game minutes accrued by each player. While coaches have presumed withdrawing high-minute players from entire games may reduce injury risk and enhance performance, a tactic which is often seen nearing the conclusion of the regular season, data to support this approach is lacking. In fact, existing data revealed the average minutes played per game did not influence on-court performance or injury risk ($p < 0.001$) in 811 NBA players competing between 2000 and 2015 [8,9]. However, it should be noted these data are not reflective of performance and injury risk in players who were rested for entire games but rather are indicative of players completing reduced game minutes. Subsequently, future studies are needed to examine the consequences and confirm the efficacy of resting high-minute players for entire games in the NBA.

Scientific information about the specific demands of air travel on performance and health in professional team sports is scarce, with research existing in soccer [40] and rugby [41], which may not directly apply to the NBA. Therefore, research is needed to understand the impact of air travel on player health and game performance across the season in the NBA. Future research on the influence of air travel in NBA players should focus on the identification of causes and symptoms of travel fatigue as well as interventions to mitigate the effects of air travel on player health and performance.

4. Conclusions and Future Research

The NBA travel schedule induces misalignments in circadian rhythm that cannot be avoided. Air travel across three time zones has been reported to induce susceptibility to travel fatigue [18,29,42–44], increase injury risk [13,29,41], and reduce game performance [13,14,17,29,32]. NBA schedule-makers and teams may succeed in mitigating the negative effects of air travel from coast to coast on sleep by implementing up-to-date, evidence-based strategies applied in other professional sports, such as blue light exposure in the morning and red light exposure in the evening, in order to resynchronize the circadian rhythms of players [45]. Other strategies include the ingestion of a high-carbohydrate, low-protein meal in the evening, which may enhance serotonin production to promote drowsiness and sleep [19,46], or the ingestion of a high-protein, low-carbohydrate meal in the morning, which may increase the uptake of tyrosine and its conversion to adrenaline, which elevates arousal and promotes alertness [44,46]. However, future studies are required to evaluate the efficacy of the abovementioned strategies in NBA players.

Despite recent schedule modifications and an increased awareness of the potential negative consequences of air travel on the health and performance of NBA players, there is still a need to implement effective strategies to address issues with sleep and travel fatigue to promote greater equity across western and eastern teams. Future research exploring various aspects of regularity, duration, directions, and conditions of air travel [13] in one or multiple NBA seasons can help identify origins of fatigue in players. Consequently, a holistic approach to future research is recommended, with some potential topics of interest encompassing descriptive and intervention-style studies.

First, it is important to understand the impact of air travel on NBA players at an individual level, given that NBA players often experience time zone transitions, which have been found to increase injury risk [9,41] and hinder performance [15,19,21,40,42,47]. Considering frequent time zone transitions often disrupt the circadian rhythm in athletes [15,16,19,26,42,43], future studies may focus on the measurement of salivary melatonin onset, adrenaline concentrations, and body temperature, as these are critical biomarkers of circadian rhythm [19,48]. Measurement of these biomarkers would provide insight into how each player individually adapts to air travel throughout the NBA season. Consequently, NBA performance support staff may then apply individualized approaches to training and game preparation to combat the negative impact of air travel.

Second, examination of various ergogenic aids will provide a better understanding of practices that may enhance physiological and perceptual responses to air travel in NBA players. For instance, nutrition [49] and hydration [49] are fundamental aspects underpinning circadian rhythm. Therefore, analyzing and comparing the hormonal responses of NBA players adopting different diets may provide NBA coaches and support staff with further insight into beneficial nutritional strategies for coping with air travel in the NBA.

Third, in order to mitigate the negative impact of air travel on mood state, it is recommended that each player's psychological and psycho-sociological reactions to air travel should be monitored during the season. For instance, comprehensive psychometric questionnaires such as the Acute Recovery and Stress Scale (ARSS) [50] and the REST-Q Sport [51] have been established as logical, practical, and versatile tools to measure self-perceived travel fatigue in professional team sports [50,51]. Considering the time constraints in the NBA, shorter customized versions of these questionnaires can be completed on a daily basis [52], which have been reported to be valid and reliable in elite Australian Rules Football [53]. However, further research is necessary to provide normative standards, especially with a focus on individual interpretations, recommendations, and compliance in NBA players.

Finally, considering that skeletal muscle and connective tissues become shortened during flights and may stiffen, it is recommended for players to avoid sitting the entire trip, and instead, walk around the cabin every hour, unless they are asleep or advised not to do so by flight staff [46]. With a tentative agreement between the NBA and Delta Airlines charters, walking inside the air cabin should be attainable, as most NBA teams (27 out of 30 teams) fly with private jets of Delta Airlines (including A319s and Boeing 757-200s) with almost 50 percent more cabin space than standard planes [54]. This cabin space allows most NBA players, who possess an average stature of 6 feet and 7 inches, to have more freedom to stand erect during air travel [54]. Additionally, simple stretching exercises can be applied while in the seat or in the cabin, which could help relax muscles while increasing blood flow and delivering oxygen and other nutrients to muscles [27,46]. As a result, stretching may reduce the negative effects of air travel on flexibility and skeletal muscle recovery. Consequently, future studies are encouraged to examine the efficacy of these in-flight travel strategies in NBA players.

Author Contributions: T.H., A.T.S., V.J.D., and J.C.-G conceived and designed the review; T.H. performed the review and developed the manuscript; A.T.S., V.J.D., and J.C.-G edited the manuscript. All authors approved the final version of the manuscript.

Funding: This research received no external funding.

Conflicts of Interest: The authors declare no conflict of interest.

References

1. Sampaio, J.; McGarry, T.; Calleja-González, J.; Sáiz, S.J.; Alcázar, X.S.; Balciunas, M. Exploring game performance in the National Basketball Association using player tracking data. *PLoS ONE* **2015**, *10*, e0132894. [CrossRef] [PubMed]
2. Official NBA Statistics and Advanced Analytics. Available online: www.stats.nba.com (accessed on 15 August 2018).
3. McLean, B.D.; Strack, D.; Russell, J.; Coutts, A.J. Quantifying physical demands in the National Basketball Association (NBA): Challenges in developing best-practice models for athlete care and performance. *Int. J. Sports Physiol. Perform.* **2018**, 1–22. [CrossRef] [PubMed]
4. Wilke, J.; Niederer, D.; Vogt, L.; Banzer, W. Head coaches' attitudes towards injury prevention and use of related methods in professional basketball: A survey. *Phys. Ther. Sport* **2018**, *32*, 133–139. [CrossRef] [PubMed]
5. Lewis, M. It's a hard-knock life: Game load, fatigue, and injury risk in the National Basketball Association. *J. Athl. Train.* **2018**, *53*, 503–509. [CrossRef] [PubMed]
6. The Official Site of the NBA. Available online: www.nba.com (accessed on 15 August 2018).
7. NBA Advanced Stats and Analytics. Available online: www.nbasavant.com (accessed on 15 August 2018).
8. Belk, J.W.; Marshall, H.A.; McCarty, E.C.; Kraeutler, M.J. The effect of regular-season rest on playoff performance among players in the National Basketball Association. *Orthop. J. Sports Med.* **2017**, *5*. [CrossRef] [PubMed]
9. Teramoto, M.; Cross, C.; Cushman, D.; Maak, T.; Petron, D.; Willick, S. Game injuries in relation to game schedules in the National Basketball Association. *J. Sci. Med. Sport* **2017**, *20*, 230–235. [CrossRef] [PubMed]

10. Philbrick, J.T.; Shumate, R.; Siadaty, M.S.; Becker, D.M. Air travel and venous thromboembolism: A systematic review. *J. Gen. Intern. Med.* **2007**, *22*, 107–114. [CrossRef] [PubMed]
11. Drakos, M.C.; Domb, B.; Starkey, C.; Callahan, L.; Allen, A. Injury in the National Basketball Association: A 17-year overview. *Sports Health* **2010**, *2*, 284–290. [CrossRef] [PubMed]
12. Coste, O.; Van Beers, P.; Touitou, Y. Hypoxia-induced changes in recovery sleep, core body temperature, urinary 6-sulphatoxymelatonin and free cortisol after a simulated long-duration flight. *J. Sleep Res.* **2009**, *18*, 454–465. [CrossRef] [PubMed]
13. Reilly, T. *Ergonomics in Sport and Physical Activity: Enhancing Performance and Improving Safety*, 1st ed.; Human Kinetics: Champaign, IL, USA, 2010; pp. 75–95.
14. Roy, J.; Forest, G. Greater circadian disadvantage during evening games for the National Basketball Association (NBA), National Hockey League (NHL) and National Football League (NFL) teams travelling westward. *J. Sleep Res.* **2017**, *27*, 86–89. [CrossRef] [PubMed]
15. Leatherwood, W.E.; Dragoo, J.L. Effect of airline travel on performance: A review of the literature. *Br. J. Sports Med.* **2013**, *47*, 561–567. [CrossRef] [PubMed]
16. Forbes-Robertson, S.; Dudley, E.; Vadgama, P.; Cook, C.; Drawer, S.; Kilduff, L. Circadian disruption and remedial interventions. *Sports Med.* **2012**, *42*, 185–208. [CrossRef] [PubMed]
17. Bishop, D. The effects of travel on team performance in the Australian national netball competition. *J. Sci. Med. Sport* **2004**, *7*, 118–122. [CrossRef]
18. Samuels, C.H. Jet lag and travel fatigue: A comprehensive management plan for sport medicine physicians and high-performance support teams. *Clin. J. Sport Med.* **2012**, *22*, 268–273. [CrossRef] [PubMed]
19. Manfredini, R.; Manfredini, F.; Fersini, C.; Conconi, F. Circadian rhythms, athletic performance, and jet lag. *Br. J. Sports Med.* **1998**, *32*, 101–106. [CrossRef] [PubMed]
20. Moore, S.; Scott, J. Beware thin air: Altitude's influence on NBA game outcomes. *JUR* **2013**, *4*, 11–17.
21. Steenland, K.; Deddens, J.A. Effect of travel and rest on performance of professional basketball players. *Sleep* **1997**, *20*, 366–369. [PubMed]
22. Nédélec, M.; McCall, A.; Carling, C.; Legall, F.; Berthoin, S.; Dupont, G. Recovery in soccer. *Sports Med.* **2013**, *43*, 9–22. [CrossRef] [PubMed]
23. Palmer, B.F. Physiology and pathophysiology with ascent to altitude. *Am. J. Med. Sci.* **2010**, *340*, 69–77. [CrossRef] [PubMed]
24. Humphreys, S.; Deyermond, R.; Bali, I.; Stevenson, M.; Fee, J.P. The effect of high altitude commercial air travel on oxygen saturation. *Anaesthesia* **2005**, *60*, 458–460. [CrossRef] [PubMed]
25. Lindgren, T. Cabin Air Quality in Commercial Aircraft. Ph.D. Thesis, Uppsala University, Uppsala, Sweden, 2003.
26. Reilly, T.; Edwards, B. Altered sleep–wake cycles and physical performance in athletes. *Physiol. Behav.* **2007**, *90*, 274–284. [CrossRef] [PubMed]
27. Hoffman, J.R.; Im, J.; Rundell, K.W.; Kang, J.; Nioka, S.; Spiering, B.A.; Kime, R.; Chance, B. Effect of muscle oxygenation during resistance exercise on anabolic hormone response. *Med. Sci. Sport Exerc.* **2003**, *35*, 1929–1934. [CrossRef] [PubMed]
28. Kraemer, W.J.; Hooper, D.R.; Kupchak, B.R.; Saenz, C.; Brown, L.E.; Vingren, J.L.; Hui Ying, L.; DuPont, W.H.; Szivak, T.K.; Flanagan, S.D.; et al. The effects of a roundtrip trans-American jet travel on physiological stress, neuromuscular performance, and recovery. *J. Appl. Physiol.* **2016**, *121*, 438–448. [CrossRef] [PubMed]
29. Youngstedt, S.D.; O'connor, P.J. The influence of air travel on athletic performance. *Sports Med.* **1999**, *28*, 197–207. [CrossRef] [PubMed]
30. Reilly, T.; Waterhouse, J. Sports performance: Is there evidence that the body clock plays a role? *Eur. J. Appl. Physiol.* **2009**, *106*, 321–332. [CrossRef] [PubMed]
31. Reilly, T.; Waterhouse, J.; Edwards, B. Jet lag and air travel: Implications for performance. *Clin. Sports Med.* **2005**, *24*, 367–380. [CrossRef] [PubMed]
32. Pollard, R.; Gómez, M.A. Components of home advantage in 157 national soccer leagues worldwide. *Int. J. Sport Exerc. Psychol.* **2014**, *12*, 218–233. [CrossRef]
33. Goumas, C. Home advantage in Australian soccer. *J. Sci. Med. Sport* **2014**, *17*, 119–123. [CrossRef] [PubMed]
34. Sack, R.L. Jet lag. *N. Engl. J. Med.* **2010**, *362*, 440–447. [CrossRef] [PubMed]
35. Entine, O.A.; Small, D.S. The role of rest in the NBA home-court advantage. *J. Quant. Anal. Sports* **2008**, *4*. [CrossRef]

36. The Gatorade Sports Science Institute. Available online: www.gssiweb.org (accessed on 15 August 2018).

37. Montgomery, P.G.; Pyne, D.B.; Hopkins, W.G.; Dorman, J.C.; Cook, K.; Minahan, C.L. The effect of recovery strategies on physical performance and cumulative fatigue in competitive basketball. *J. Sports Sci.* **2008**, *26*, 1135–1145. [CrossRef] [PubMed]

38. Delextrat, A.; Calleja-González, J.; Hippocrate, A.; Clarke, N.D. Effects of sports massage and intermittent cold-water immersion on recovery from matches by basketball players. *J. Sports Sci.* **2013**, *31*, 11–19. [CrossRef] [PubMed]

39. Delextrat, A.; Hippocrate, A.; Leddington-Wright, S.; Clarke, N.D. Including stretches to a massage routine improves recovery from official matches in basketball players. *J. Strength Cond. Res.* **2014**, *28*, 716–727. [CrossRef] [PubMed]

40. Fowler, P.M.; McCall, A.; Jones, M.; Duffield, R. Effects of long-haul transmeridian travel on player preparedness: Case study of a national team at the 2014 FIFA World Cup. *J. Sci. Med. Sport* **2017**, *20*, 322–327. [CrossRef] [PubMed]

41. Fuller, C.W.; Taylor, A.E.; Raftery, M. Does long-distance air travel associated with the Sevens World Series increase players' risk of injury? *Br. J. Sports Med.* **2015**, *49*, 458–464. [CrossRef] [PubMed]

42. Fowler, P.M.; Knez, W.; Crowcroft, S.; Mendham, A.E.; Miller, J.; Sargent, C.; Duffield, R. Greater effect of east vs. west travel on jet-lag, sleep and team-sport performance. *Med. Sci. Sports Exerc.* **2017**, *49*, 2548–2561. [CrossRef] [PubMed]

43. Thornton, H.R.; Miller, J.; Taylor, L.; Sargent, C.; Lastella, M.; Fowler, P.M. Impact of short-compared to long-haul international travel on the sleep and wellbeing of national wheelchair basketball athletes. *J. Sports Sci.* **2017**, *36*, 1476–1484. [CrossRef] [PubMed]

44. Leathwood, P. Circadian rhythms of plasma amino acids, brain neurotransmitters and behaviour. In *Biological Rhythms in Clinical Practice*, 1st ed.; Arendt, J., Minors, D., Waterhouse, J., Eds.; Butterworths: London, UK, 1989; pp. 136–159.

45. Czeisler, C.A.; Allan, J.S.; Strogatz, S.H. Bright light resets the human circadian pacemaker independent of the timing of the sleep-wake cycle. *Science* **1986**, *233*, 667–671. [CrossRef] [PubMed]

46. Meir, R. Managing transmeridian travel: Guidelines for minimizing the negative impact of international travel on performance. *Strength Cond. J.* **2002**, *24*, 28–34. [CrossRef]

47. Srinivasan, V.; Singh, J.; Pandi-Perumal, S.R.; Brown, G.M.; Spence, D.W.; Cardinali, D.P. Jet lag, circadian rhythm sleep disturbances, and depression: The role of melatonin and its analogs. *Adv. Ther.* **2010**, *27*, 796–813. [CrossRef] [PubMed]

48. Roach, G.D.; Rogers, M.; Dawson, D. Circadian adaptation of aircrew to transmeridian flight. *Aviat. Space Environ. Med.* **2002**, *73*, 1153–1160. [PubMed]

49. Halson, S.L. Sleep in elite athletes and nutritional interventions to enhance sleep. *Sports Med.* **2014**, *44*, 13–23. [CrossRef] [PubMed]

50. Kölling, S.; Hitzschke, B.; Holst, T.; Ferrauti, A.; Meyer, T.; Pfeiffer, M.; Kellmann, M. Validity of the acute recovery and stress scale: Training monitoring of the German junior national field hockey team. *Int. J. Sports Sci. Coach.* **2015**, *10*, 529–542. [CrossRef]

51. Bresciani, G.; Cuevas, M.J.; Garatachea, N.; Molinero, O.; Almar, M.; De Paz, J.A.; Márquez, S.; González-Gallego, J. Monitoring biological and psychological measures throughout an entire season in male handball players. *Eur. J. Sports Sci.* **2010**, *10*, 377–384. [CrossRef]

52. Gastin, P.B.; Meyer, D.; Robinson, D. Perceptions of wellness to monitor adaptive responses to training and competition in elite Australian football. *J. Strength Cond. Res.* **2013**, *27*, 2518–2526. [CrossRef] [PubMed]

53. Taylor, K.; Chapman, D.; Cronin, J.; Newton, M.J.; Gill, N. Fatigue monitoring in high performance sport: A survey of current trends. *J. Aust. Strength Cond.* **2012**, *20*, 12–23.

54. NBA Players Get Roomier Chartered Jets as Delta Air Adds Teams. Available online: https://www.bloomberg.com/news/articles/2015-07-06/nba-players-get-roomier-chartered-jets-as-delta-air-adds-teams (accessed on 28 June 2018).

© 2018 by the authors. Licensee MDPI, Basel, Switzerland. This article is an open access article distributed under the terms and conditions of the Creative Commons Attribution (CC BY) license (http://creativecommons.org/licenses/by/4.0/).

sports

MDPI

Article

Anthropometric Variables and Somatotype of Young and Professional Male Basketball Players

Karol Gryko [1] **, Anna Kopiczko** [2] **, Kazimierz Mikołajec** [3] **, Petr Stasny** [4,*] **and Martin Musalek** [4]

[1] Department of Athletics and Team Sport Games, Józef Piłsudski University of Physical Education, Warszawa 00-968, Poland; gryczan@wp.pl
[2] Department of Anthropology and Health Promotion, Józef Piłsudski University of Physical Education, Warszawa 00-968, Poland; anna.kopiczko@awf.edu.pl
[3] Department of Theory and Practice of Sport, The Jerzy Kukuczka Academy of Physical Education, Katowice 40-065, Poland; k.mikolajec@awf.katowice.pl
[4] Faculty of Physical Education and Sport, Charles University, 162 52 Prague, Czech Republic; musalek@ftvs.cuni.cz
* Correspondence: stastny@ftvs.cuni.cz; Tel.: +420-777-198-764

Received: 15 December 2017; Accepted: 22 January 2018; Published: 29 January 2018

Abstract: Background: Determining somatic models and profiles in young athletes has recently become a fundamental element in selecting basketball playing positions. The aim of this study was to assess the relationship between the body build of young and adult elite male basketball players at different playing positions. Methods: Participants consisted of 35 young (age: 14.09 ± 0.30 years, $n = 35$) and 35 adult professional basketball players (age: 24.45 ± 5.40 years, $n = 35$) competing in elite leagues. The anthropometric characteristics assessed included body mass, body height, skinfolds, somatotypes, girths, and breadths. Results: The centers in both age groups were significantly taller and heavier ($p < 0.001$) compared to forwards and guards. The greatest difference between categories were in the guards' personal height (from 169.36 to 186.68 = 17.32 cm). The guards from the professional team were closest in height to the forwards (difference = 7.17 cm) compared to young players where the difference between guards and forwards was 13.23 cm. Young competitors were more ectomorphic (2.12-3.75-4.17), while professional players were more mesomorphic (2.26-4.57-3.04). Significant criteria for center selection at professional level seems to be personal height and arm span ratio. Conclusions: The results indicate that the selection for basketball playing positions should include the analysis of body height and mass, shoulder breadth, humerus breadth, femur breadth and specifically for centers the difference between personal the height and arm span.

Keywords: maturation; elite sport; playing position; body composition; youth athletes; talent selection

1. Introduction

Performance in basketball depends on many factors, with the most important one being players' somatic build, as well as technical, tactical, motor, physiological, and psychological preparation. A basketball coach must supervise balanced development of players, i.e., physique, visual and motor coordination improvement and development of necessary motor abilities, considering evolutionary processes connected with the pace of growth and maturation of players [1–3]. In basketball, an individualized approach and making anthropometric diagnoses are basic elements of the selection process and of developing a long-term sports career.

Anthropometric measurements, determination of somatic build models, and somatic profiles have recently become fundamental research areas for sports training specialists [4–8]. Somatic profiles of basketball players have been widely recognized as a crucial factor in the selection process and as a performance predictor [5,9–11]. Anthropometric characteristics, such as body fat, skinfold thickness, body height, arm span, and body circumferences, were determined to be principal components in elite basketball players; therefore, they are often regarded as indicators of the level of play [8].

Previous analyses of somatic characteristics in basketball players indicate that body measurements are essential in the general selection process and in assigning playing positions [12]. Moreover, somatic parameters have an impact on players performance in condition tests [13]. Tests on both young and professional players revealed that individuals who were taller in stature, had more mesomorph component, and had longer limbs obtained higher scores regarding efficiency on the court and achieved better physiological parameters [14]. The crucial component in the process of assigning specific playing positions is body height [4], in which the tallest players are selected as centers (close to the basket), and those of shorter stature as guards (on the perimeter, further away from the basket) [5,15]. Additionally, the competitors playing in different positions also revealed differences in body girths (thigh, calf, arm, and forearm girths) between players [16].

Somatotype, defined as the description of such morphological components as endomorph, mesomorph and ectomorph, is another valuable tool for the accurate assessment of somatic parameters needed in a given sport [17]. Popovic et al. [18] observed that male basketball players are likely to display a mesomorph somatotype, but there are also professional players from top teams with mixed and balanced somatotypes. Moreover, the somatotype and other anthropometric variables might be specific to geographical region, especially during growth and maturation [19].

Considering the current state of knowledge in this field, it might be beneficial to examine breadth- and circumference-related aspects of body build. Moreover, there is lack of previous studies comparing the anthropometrics in young and senior elite basketball players. Therefore, the aim of this study was to compare body fat, length parameters, girths, circumferences, somatotypes and breadth-related measurements between players of different positions on young and adult male elite basketball teams. Furthermore, this study examined the relationship or specificity in selected anthropometric characteristics and basketball playing position.

2. Materials and Methods

Anthropometric measurements were assessed by experienced anthropometric technician in optimal climatic conditions in accordance with standards set by the International Society for the Advancement of Kinanthropometry (ISAK) [20]. The following variables were measured: age; basketball experience; body mass, body height and arm span (GPM anthropometer, Siber Hegner, Zurich, Switzerland); relaxed arm girths; flexed arm girth; calf girths (Holtain anthropometric tape, Crymych, UK); shoulder breadth; humerus breadth; femur breadth (GPM big and small spreading caliper, Zurich, Switzerland); and skinfolds from the triceps, subscapular, biceps, iliac crest, supraspinal, abdominal, and medial calf (Harpenden Skinfold Caliper, British Indicators, West Sussex, UK).

Body mass and body fat (BF) percentages were determined with the Bioelectrical Impedance Analysis (BIA) using the Tanita BC-418 device (Amsterdam, The Netherlands). Somatotype was calculated according to the Heath-Carter method [21] using the Somatotype 1.2.6 computer program (MER Goulding Software Development, Geeveston, Australia).

The study included 70 male basketball players from two different age categories (young and adult, Table 1). The first group consisted of young elite basketball players (*n* = 35) from the Mazovia regional team (age: 14.09 ± 0.30 years) that qualified for the 2014–2016 Polish Championships of Regional Teams. The team members are the best players selected from sports clubs in the Mazovia Region who are medalists from the Polish Youth Championships. The second group (*n* = 35) consisted of professional adult basketball players (age: 24.45 ± 5.40 years) competing in the highest-level league in Poland. The playing position of the players was determined by their common match nomination in regional team or professional club.

Table 1. Characteristics of young and professional male basketball players and differences between both groups.

Variable	Young Players (n = 35)		Adult Professional Players (n = 35)	
	Mean ± SD	Range	Mean ± SD	Range
Age (years)	14.09 ± 0.30	13.37–14.47	24.45 ± 5.40	18.45–36.83
Basketball experience (years)	3.73 ± 1.24	1.0–7.0	13.43 ± 3.53	8.0–22.0
Body mass (kg)	64.98 ± 10.70 [†]	43.6–99.7	90.23 ± 10.50	72.6–116.6
Body height (cm)	179.22 ± 8.41 [†]	158.8–194.1	193.44 ± 8.07	174.3–219.0
Arm span (cm)	183.09 ± 8.15 [†]	165.5–196.6	197.78 ± 9.17	179.1–223.0
Body mass index (BMI)	20.12 ± 2.16 [†]	17.0–28.2	24.0 ± 1.81	19.9–28.1
Body fat (%)	11.0 ± 3.79 [†]	4.3–22.8	14.01 ± 3.06	8.7–22.60
Triceps skinfold (mm)	8.64 ± 3.23	5.0–24.0	7.56 ± 2.38	3.3–13.3
Subscapular skinfold (mm)	7.51 ± 3.03 [†]	5.0–23.2	10.18 ± 2.15	6.5–18.2
Biceps skinfold (mm)	4.46 ± 1.59	3.0–10.0	4.78 ± 1.64	3.07–9.73
Iliac crest skinfold (mm)	10.18 ± 4.52	6.2–25.5	11.79 ± 4.76	6.23–27.2
Supraspinal skinfold (mm)	7.02 ± 3.12	4.1–15.5	8.30 ± 2.45	4.7–14.9
Abdominal skinfold (mm)	10.2 ± 5.98	4.3–29.0	9.91 ± 4.81	5.0–28.3
Medial calf skinfold (mm)	9.07 ± 4.08	4.5–26.0	7.64 ± 2.96	3.53–17.0
Relaxed arm girth (cm)	25.60 ± 2.36 [†]	20.5–32.0	31.37 ± 2.03	28.2–36.0
Flexed arm girth (cm)	27.89 ± 2.38 [†]	22.4–34.4	34.80 ± 2.23	30.2–38.0
Calf girth (cm)	35.73 ± 2.83 [†]	30.0–46.0	39.63 ± 2.45	35.0–45.0
Shoulder breadth (cm)	39.04 ± 1.83 [†]	35.4–43.1	42.90 ± 1.72	39.8–47.2
Humerus breadth (cm)	7.09 ± 0.43 [†]	6.0–8.0	7.59 ± 0.51	6.7–8.9
Femur breadth (cm)	9.91 ± 0.43 [*]	9.0–10.7	10.39 ± 0.84	9.0–12.00
Endomorphy	2.12 ± 0.81	1.16–5.57	2.26 ± 0.59	1.19–3.66
Mesomorphy	3.75 ± 1.01 [*]	1.23–5.88	4.57 ± 1.07	2.31–6.95
Ectomorphy	4.17 ± 1.08 [†]	1.18–6.36	3.04 ± 0.89	1.22–5.38

[*] Significantly different from adult professional basketball players ($p < 0.01$); [†] Significantly different from adult professional basketball players ($p < 0.001$).

Prior to the commencement of the study, all of the participants were informed about the study's aims and conduct, as well as about the possibility of resigning from research participation without providing any causes at any time. An informed consent provided by the participants or their legal representatives signature (if age below 18 years) was the study inclusion criterion, whereas contraindications for being subjected to anthropometric measurement procedures or bioelectrical impedance analysis were the exclusion criteria. The research was conducted in accordance with approval from the Ethics Committee for Scientific Research of the University of Physical Education in Warsaw, and the study was completed according to the rules and regulations of the Declaration of Helsinki [22].

All statistical analyses were performed in STATISTICA version 12 (StatSoft, Inc., Tulsa, OK, USA). The means, standard deviations (SD), and maximum and minimum values were used for group descriptions (Table 1), and somatotype values were expressed in a somatochart for both groups (Figure 1). The Shapiro–Wilk test was applied to examine the data normality distribution. One-way ANOVA (post-hoc Tukey tests, for equal sample sizes, with $p < 0.01$, Hays $\omega^2 < 0.08$ considered significant) was employed to assess the significance of differences in values referring to anthropometric and somatic features between young and professional groups of basketball players. The MANOVA (post-hoc Tukey tests, for unequal sample sizes, with $p < 0.05$, Hays $\omega^2 < 0.09$, considered significant) was used to show significant differences in parameters describing young and professional basketball players in respective positions. If appropriate, the Kruskal–Wallis non-parametric ANOVA was used for selected parameters. For ANOVA analyses, the players were divided into three groups according to playing positions: guards, forwards, and centers.

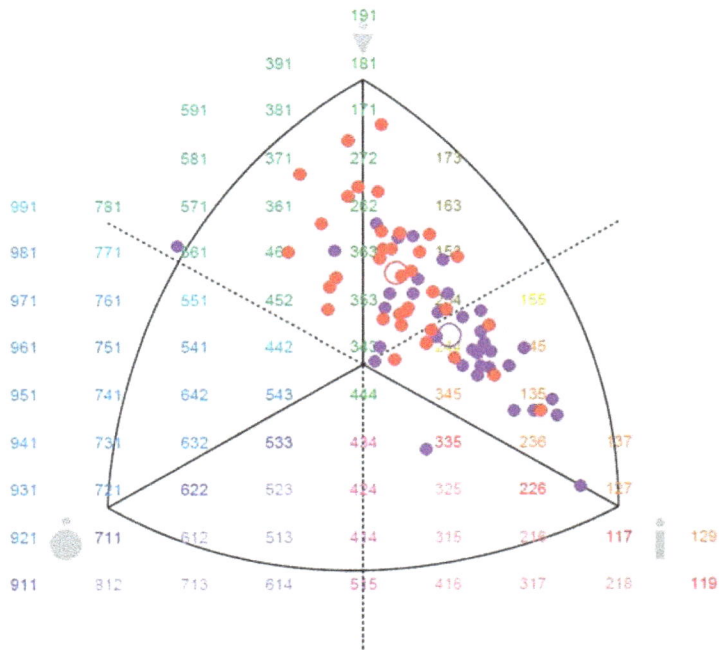

Figure 1. Somatochart of the examined basketball players. The circle is the mean profile of each group: • Adult professional basketball players, • Young basketball players.

To determine the strength of the association between playing positions and anthropometric variables (nominal-by-interval variable), the Hay ω was used. Hay ω from 0.10–0.30 was thought to represent a weak association; coefficient from 0.30–0.50 was considered a moderate association; and coefficient of and greater than 0.50 was considered a strong association [23]. Effect size was calculated as Cohen f (Small effect: <0.10; medium effect: 0.10–0.40; large effect: <0.40) [24,25]. Significant differences and correlations was assumed at $p < 0.05$.

3. Results

The Shapiro-Wilk test revealed no grounds for rejecting the hypothesis of normality in both groups without specification of playing position, and further in body height, body mass, fat percentage, breadth parameters and somototypes if considering players position. Anthropometric characteristics of young and adult male players revealed that young players demonstrated significantly ($p < 0.001$) lower values of body height, arm span, body mass, (body mass index) BMI, body fat (BF), shoulder breadth, humerus breadth and femur breadth ($p < 0.01$), see Table 1. Furthermore, younger players had significantly ($p < 0.001$) lower values of girth parameters: relaxed arm girth, flexed arm girth, calf girth and subscapular skinfold (Table 1). In addition, a significantly ($p < 0.01$) lower percentage of the mesomorphic component and higher percentage of the ectomorphic component were noted in young basketball players (Table 1 and Figure 1). Young as well as adult basketball players had wider arm span than personal height.

When considering the absolute differences in personal height between young and professional adult male basketball players regarding to players position, we revealed that increments were not proportionate. The greatest difference in personal height was identified in guards (from 169.36 to 186.68 cm, a difference of 17.32 cm). Guards and forwards from the adult professional team were close in height (difference = 7.17 cm), whereas guards and forwards on the young team were different in height (difference = 13.23 cm). In conformity with results regarding personal height, in young players, the arm span of 146 cm in centers was significantly greater ($F_{2,33} = 32.89$, $p < 0.001$, $\omega^2 = 0.23$) compared to guards and forwards (147 cm), and guards had the shortest arm span. Adult centers had a significantly wider arm span ($F_{2,33} = 22.26$, $p < 0.01$, Hays $\omega^2 = 0.19$) compared to guards and forwards. Professional players centers had significantly greater difference between personal height and arm span ($F_{2,33} = 3.89$, $p < 0.05$, $\omega^2 = 0.14$) compared to guards and forwards, which seems to be a significant factor for selection of center position: average difference between personal height and arm span: center = 6.86 cm; guards = 3.56 cm; forwards = 2.42 cm, respectively.

Guards in both categories had significantly narrower breadth of humerus epicondyle (young players, $F_{2,33} = 24.12$, $p < 0.001$, $\omega^2 = 0.19$; professional players: $F_{2,33} = 13.22$, $p < 0.001$, $\omega^2 = 0.15$) and femur epicondyle (young players, $F_{2,33} = 45.12$, $p < 0.001$, $\omega^2 = 0.21$; professional players, $F_{2,33} = 14.78$, $p < 0.001$, $\omega^2 = 0.17$) than centers. Additionally, young guards had lower shoulder breadth values ($F_{2,33} = 130.9$, $p < 0.001$, $\omega^2 = 0.48$) compared to forwards and centers (Table 2). The endomorphic component was greater in the centers than in the forwards ($F_{2,33} = 26.12$, $p < 0.001$, $\omega^2 = 0.21$) among young players. The ectomorphic component was significantly ($F_{2,33} = 34.92$, $p < 0.001$, $\omega^2 = 0.27$) more prevalent in the forwards than in the guards in young players (Table 2).

Based on anthropometric measurements, three somatic types were determined in basketball players from both groups. The mean somatotype of young players was characterized by the following code: 2.12-3.75-4.17, indicating that the average young player had an ecto-mesomorphic body build. The mean somatotype of professional players was defined by the following code: 2.26-4.57-3.04, meaning that the adult players were more meso-ectomorphic (Figure 1).

Since a portion of measured skinfolds expressed non-parametric characteristics, we used Kruskal-Wallis non-parametric ANOVA to analyze these cases. In young players, there were clear differences between players of different positions in three front trunk skinfolds (iliac, supraspinal, and abdominal) and in triceps and medial calf skinfolds. Centers had significantly greater skinfold measurements than forwards in triceps (Chi-Square $H_{2,31} = 8.75$, $p = 0.03$), calf ($H_{2,31} = 7.94$, $p = 0.02$), iliac ($H_{2,31} = 6.70$, $p = 0.01$), supraspinal ($H_{2,31} = 5.71$, $p = 0.01$), and abdominal ($H_{2,31} = 4.32$, $p = 0.03$). On the other hand, for professional players, there were no significant differences in skinfold values among different player positions.

Table 3 illustrates the correlations for the calculated anthropometric indices and playing positions (guards, forwards, centers) for each of the groups and as a combined dataset. In the group of young basketball players, significant strong positive correlations were found in body height (with large effect size), body mass, and shoulder breadth, while significant moderate positive correlations were noted between for all other parameters except arm span.

For professional basketball players, there were significant positive correlations between body mass (with large effect size), body height, flexed arm girth and playing position. Moreover, moderate positive correlations to playing position were noted in arm span, calf girth, relaxed arm girth, shoulder breath, humerus breath, femur breath, and BMI.

Table 2. Baseline characteristics of young and professional basketball male players in relation to their playing positions.

Variable	Guards		Forwards		Centers	
	Young (n = 12)	Adult Professional (n = 12)	Young (n = 11)	Adult Professional (n = 11)	Young (n = 12)	Adult Professional (n = 12)
Age	14.14 ± 0.31	23.94 ± 5.17	14.03 ± 0.33	24.37 ± 5.13	14.10 ± 0.28	25.04 ± 6.24
Body mass (kg)	C 57.10 ± 6.64 †	C 81.08 ± 4.61	C 63.71 ± 6.67 †	C 89.25 ± 8.55	G,F 74.02 ± 10.51 †	G,F 100.29 ± 7.10
Body height (cm)	C 169.36 ± 5.16 †	C 186.68 ± 5.9	C 182.59 ± 3.81 †	C 193.85 ± 4.39 †	G,F 185.98 ± 3.39 †	G,F 199.83 ± 7.37
Arm span (cm)	C,F 174.44 ± 5.98 †	C 190.23 ± 4.52	G 184.90 ± 3.11 †	C 196.28 ± 5.82	G 190.08 ± 4.76 †	G,F 206.70 ± 7.58
Body mass index	19.88 ± 1.51 †	23.06 ± 0.86	19.11 ± 1.46 †	23.75 ± 2.09	21.28 ± 2.78 †	25.16 ± 1.72
Body fat (%)	11.07 ± 3.18	13.21 ± 1.93	9.53 ± 3.09	13.28 ± 3.13	12.27 ± 4.65	15.47 ± 3.58
Triceps SF (mm)	8.29 ± 1.04	8.01 ± 2.58	C 7.17 ± 2.02	6.48 ± 1.8	F 10.33 ± 4.7	8.1 ± 2.49
Subscapular SF (mm)	6.88 ± 1.45	9.82 ± 1.46	C 6.61 ± 1.07	9.51 ± 2.08	F 8.97 ± 4.64	11.14 ± 2.57
Biceps SF (mm)	4.21 ± 1.0	5.31 ± 2.35	3.85 ± 1.07	4.44 ± 1.26	5.27 ± 2.15	4.58 ± 0.96
Iliac skinfold (mm)	8.94 ± 2.63	12.16 ± 4.5	C 8.06 ± 1.12	10.73 ± 4.03	F 3.37 ± 6.16	12.38 ± 5.77
Supraspinal SF (mm)	6.25 ± 1.99	8.11 ± 2.46	C 5.48 ± 1.3	8.34 ± 2.02	F 9.21 ± 4.05	8.46 ± 2.96
Abdominal SF (mm)	9.22 ± 4.78	9.43 ± 4.4	C 7.27 ± 1.43	9.28 ± 3.18	F 13.87 ± 7.84	10.96 ± 6.42
Medial calf SF (mm)	8.37 ± 2.18	7.31 ± 2.47	C 6.95 ± 2.03	7.06 ± 2.73	F 11.72 ± 5.5	8.51 ± 3.6
Relaxed arm girth (cm)	24.82 ± 1.81 †	30.54 ± 1.64	25.04 ± 2.34 †	30.64 ± 2.5	26.88 ± 2.49 †	32.88 ± 0.79
Flexed arm girth (cm)	27.25 ± 2.22 †	34.04 ± 1.67	27.0 ± 1.7 †	C 33.68 ± 2.62	29.35 ± 2.53 †	F,G 36.57 ± 1.0
Calf girth (cm)	34.38 ± 1.84 †	38.07 ± 1.86	35.74 ± 1.89 †	39.8 ± 2.71	37.09 ± 3.75 †	F 41.04 ± 1.86
Shoulder breadth (cm)	C,F 37.35 ± 1.14 †	41.89 ± 1.48	G 39.38 ± 1.46 †	42.69 ± 1.15	G 40.43 ± 1.33 †	44.09 ± 1.75
Humerus breadth (cm)	C 6.83 ± 0.37 †	C 7.36 ± 0.42	7.07 ± 0.37	7.54 ± 0.41	G 7.36 ± 0.39 *	G 7.88 ± 0.56 *
Femur breadth (cm)	C 9.65 ± 0.41	C 9.98 ± 0.66	9.97 ± 0.42	10.7 ± 0.82	G 10.12 ± 0.35	G 10.52 ± 0.92
Endomorphy	2.09 ± 0.42	2.34 ± 0.49	C 1.67 ± 0.37	2.07 ± 0.54	F 2.56 ± 1.14	2.34 ± 0.72
Mesomorphy	4.34 ± 0.68	4.62 ± 0.9	3.23 ± 0.83 *	4.51 ± 1.42	3.64 ± 1.17	4.59 ± 0.94
Ectomorphy	F 3.68 ± 0.7	3.0 ± 0.66	G 4.94 ± 0.81 *	3.22 ± 1.06	3.96 ± 1.26	2.92 ± 0.96

† Significantly different between young and adult professional at $p < 0.01$, * Significantly different between young and adult professional at $p < 0.05$, C Significantly different from centers in the same age group at $p < 0.05$, F Significantly different from forwards in the same age group at $p < 0.05$, G Significantly different between guards in the same age group at $p < 0.05$, SF = skinfold.

Table 3. The strength of association for the calculated anthropometric variables and playing positions (guards, forwards, centers) for each group of male basketball players as a combined dataset.

Variable	Young Players (n = 35)	Effect Size Young	Adult Players (n = 35)	Effect Size Adult
Body mass (kg)	0.62 **	0.31 †	0.75 **	0.64 ††
Body height (cm)	0.86 **	1.42 ††	0.66 **	0.38 †
Arm span	0.17	0.02	0.38 *	0.09
Triceps skinfold (mm)	0.33 *	0.06	0.2	0.02
Subscapular skinfold (mm)	0.25	0.03	0.25	0.03
Biceps skinfold (mm)	0.31 *	0.06	0.21	0.2
Iliac crest skinfold (mm)	0.47 *	0.14 †	0.02	<0.01
Supraspinal skinfold (mm)	0.47 *	0.14 †	0.09	<0.01
Abdominal skinfold (mm)	0.41 *	0.10	0.09	<0.01
Medial calf skinfold (mm)	0.44 *	0.12 †	0.10	<0.01
Relaxed arm girth (cm)	0.32 *	0.06	0.49 *	0.15 †
Flexed arm girth (cm)	0.39 *	0.09	0.54 **	0.20 †
Calf girth (cm)	0.32 *	0.06	0.46 *	0.13 †
Shoulder breadth (cm)	0.69 **	0.45 ††	0.49 *	0.15 †
Humerus breadth (cm)	0.46 *	0.13 †	0.36 *	0.07
Femur breadth (cm)	0.41 *	0.10	0.32 *	0.04
Endomorphy	0.4 *	0.10	0.03	<0.01
Mesomorphy	0.4 *	0.10	0.02	<0.01
Ectomorphy	0.46 *	0.13	0.09	<0.01
Body mass index (BMI)	0.35 *	0.06	0.44 *	0.12 †
Body Fat (%)	0.35 *	0.06	0.37	0.08

* Moderate association: 0.30–0.50; ** strong association: >0.50; † medium effect size: 0.10–0.40; †† strong effect size: >0.40.

4. Discussion

The main finding of this study is that the adult male professional players are more similar regarding playing position in somatic parameters than young male players. The weight and body height are the main selective parameters in both young and professional players, however the strength of this relationship is decreased in professional players for body height. The centers are taller and heavier than forward and guards, while basic somatic features, such as body mass and height, in the examined young and professional basketball players were similar to those observed in previous studies on elite players in Poland and abroad [12,26]. Our results showed that strong correlations were found between body height, body mass, shoulder breadth, and playing positions in young players. For adult players, there were strong correlations between playing positions and body mass, body height, and flexed arm girth.

Although young male players had most of anthropometric results associated with playing position (e.g., height), there was no significant associating between playing position and arm span (Table 3). On the other hand, professional players had moderate association between playing position and arm span. Because the arm span was previously associated with professional draft status [11,27] and professional centers have the largest arm span, the young players with largest arm span should be preferred on the centers position.

The present study revealed significantly smaller basic body parameters, such as body mass and height, in young male players. Analogous differences were found in breadth-related skeleton features (significantly lower values of arm span, shoulder breadth, humerus breadth and femur breadth were noted in young basketball players). These findings might amend the knowledge about ontogenetic variability of morphological and structural capabilities of basketball players in various stages of training. A previous study [28] reported the differences in body build between young (similar age as our group) and cadet age in measurements of epiphysis diameters, body fat, tibia length, femur and trunk length, while indicating the importance of proximal bone parts development.

Our findings can be referred to the results obtained by Abdelkrim [12], where Tunisian male basketball players demonstrated mean body heights between 192.0 cm and 198.4 cm (U-18: 192.0 ± 7.3 cm; U-20: 199.2 ± 7.3 cm; seniors: 198.4 ± 6.2 cm). Their mean body mass was between

83.7 kg and 91.5 kg (U-18: 83.7 ± 8.2 kg; U-20: 91.4 ± 8.3 kg; seniors: 91.5 ± 7.2 kg). Our research revealed greater differences in body height and mass in both groups, which might be a result of low age in our young group. Similar to the research on Tunisian basketball players [12] and Spanish players from different professional leagues [8], our investigation showed that, regardless of age, the centers were the tallest of all the players. These similar studies [8,12] and our study revealed that the mean body height of the centers was almost 200 cm. The centers also had the highest body mass compared to players from other positions. Another study conducted at the first National Collegiate Athletic Association (NCAA) male division showed higher centers (205.5 ± 6 cm) and forwards (198 ± 3.8) that our study [29]. The young centers in our study were smaller and had a shorted arm span than Australian plyers (U-16: height 195 ± 4 cm; arm span 199 ± 5 cm) [10].

A previous study [30] observed that performance outcomes like agility or vertical jumps in elite male players were not related to body fat content in basketball. Our study revealed that the centers from the young male group exhibited the highest values of measured skinfolds compared to the forwards. Notably, compared to forwards, the young centers demonstrated the highest values of abdominal, triceps, subscapular, suprailiacal and calf folds. Since the body fat and body fat distribution in our study were not associated with professional playing position, we suggest avoiding selecting young players to playing positions according to those parameters.

The distribution of somatotypes in the group of young male players mainly covered the area of the somatogram close to ectomorphy, with the exception of two extreme cases (extreme ectomorph and endo-mesomorph). In adult professional players, the distribution close to mesomorphy prevailed. Our findings can be compared to a study by Martinez [31], who assessed somatotype profiles of Mexican Professional Basketball League players aged approximately 25 years. The mean value of endomorphic, mesomorphic, and ectomorphic components in that study was 2.94, 6.35, and 2.06, respectively. Our study showed that mean somatotype of professional players differed from that of Mexican competitors, i.e., the former group displayed lower values of endomorphic and mesomorphic components and higher values of ectomorphy than the latter. Regardless of their playing positions, Mexican competitors manifested extreme mesomorphic profiles, whereas professional players from our study were meso-ectomorphic what is more typical somatotype profile in elite collective sports [21,32,33].

Basketball training coaches claim that tasks performed by centers are fundamental in terms of offensive and defensive actions. Therefore, in elite teams, centers have specific body parameters that correlate with their roles on the court [7,34]. However, as a complex team sport, basketball requires proper a coach to not only train a professional team, but first and foremost, to identify and select children in the process of training. Thus, it is necessary to conduct further longitudinal research to determine useful body build characteristics and somatotypes in basketball players.

One of the study's limitation is the absence of biological age of young male players, where the greatest growth of Polish boys and greatest differences in biological age were observed between 12 and 15 years old [35]. The estimation of maturity offset by anthropometric measure was not possible with sufficient validity and reliability due to the extreme values of Polish elite basketball players [19,36]. Therefore, we can´t avoid that elite players were included in the observed team for their earlier maturation [37,38] or another bias. Another limitation is the absence of performance values of measured players, however these data have to be collected during ongoing longitudinal and further research rather than from retrospective statistic.

5. Conclusions

The results of this study indicate that anthropometric assessment of body build, as well as somatotype analysis, may be key factors in the process of talent identification in basketball. It should be highlighted that the selection for basketball and specifically for playing positions should include the analysis of somatic build features such as body height and mass, shoulder breadth, flexed arm girth and arm span. According to specialists, basketball is a dynamic team sport and, therefore, the determination of body build profiles may become a key factor in assessing players' capabilities in regard to their fitness levels and efficiency during performance. It seems that somatic parameter differences between player positions in young male players does not play a key role in adult players. In male adults, there are somatic predispositions for centers (such as the height, weight, arm span and girths), while the body build of forwards tends to be similar to that of the centers. The position with the lowest requirement for body size is the guard. Coaches should not pay attention to the body fat and body fat distribution to select players to their playing position of young male players. On the other hand the height, weight and arm span should be considered for such selection.

Acknowledgments: This study was supported by a research grant of Charles University, Czech Republic (UNCE 032), and by the Józef Piłsudski University of Physical Education in Warsaw as part of its statutory activities (DS-279).

Author Contributions: The G.K., K.A. and M.K. conceived and designed the experiments; G.K., K.A. and M.K. performed the experiments G.K., S.P., M.M., K.A. and M.K. analyzed the data; S.P. and M.M. contributed reagents/materials/analysis tools; G.K., S.P., M.M., K.A. and M.K. wrote the paper.

Conflicts of Interest: The authors declare no conflict of interest.

References

1. Sánchez-Muñoz, C.; Zabala, M.; Williams, K. Anthropometric variables and its usage to characterise elite youth athletes. In *Handbook of Anthropometry*; Springer: Berlin, Germany, 2012; pp. 1865–1888.
2. Šimonek, J.; Horička, P.; Hianik, J. Differences in pre-planned agility and reactive agility performance in sport games. *Acta Gymnica* **2016**, *46*, 68–73. [CrossRef]
3. Hůlka, K.; Lehnert, M.; Bělka, J. Reliability and validity of a basketball-specific fatigue protocol simulating match load. *Acta Gymnica* **2017**, *47*, 92–98. [CrossRef]
4. Dežman, B.; Trninić, S.; Dizdar, D. Expert model of decision-making system for efficient orientation of basketball players to positions and roles in the game–Empirical verification. *Coll. Antropol.* **2001**, *25*, 141–152. [PubMed]
5. Ostojic, S.M.; Mazic, S.; Dikic, N. Profiling in basketball: Physical and physiological characteristics of elite players. *J. Strength Cond. Res.* **2006**, *20*, 740–744. [CrossRef] [PubMed]
6. Montgomery, P.G.; Pyne, D.B.; Hopkins, W.G.; Dorman, J.C.; Cook, K.; Minahan, C.L. The effect of recovery strategies on physical performance and cumulative fatigue in competitive basketball. *J. Sports Sci.* **2008**, *26*, 1135–1145. [CrossRef] [PubMed]
7. Sampaio, J.; Janeira, M.; Ibáñez, S.; Lorenzo, A. Discriminant analysis of game-related statistics between basketball guards, forwards and centres in three professional leagues. *Eur. J. Sport Sci.* **2006**, *6*, 173–178. [CrossRef]
8. Vaquera, A.; Santos, S.; Villa, J.G.; Morante, J.C.; García-Tormo, V. Anthropometric characteristics of Spanish professional basketball players. *J. Hum. Kinet.* **2015**, *46*, 99–106. [CrossRef] [PubMed]
9. Bayios, I.A.; Bergeles, N.K.; Apostolidis, N.G.; Noutsos, K.S.; Koskolou, M.D. Anthropometric, body composition and somatotype differences of Greek elite female basketball, volleyball and handball players. *J. Sports Med. Phys. Fit.* **2006**, *46*, 271–280.
10. Hoare, D.G. Predicting success in junior elite basketball players—The contribution of anthropometic and physiological attributes. *J. Sci. Med. Sport* **2000**, *3*, 391–405. [CrossRef]
11. Berri, D.J.; Brook, S.L.; Fenn, A.J. From college to the pros: Predicting the NBA amateur player draft. *J. Prod. Anal.* **2010**, *35*, 25–35. [CrossRef]
12. Ben Abdelkrim, N.B.; Chaouachi, A.; Chamari, K.; Chtara, M.; Castagna, C. Positional role and competitive-level differences in elite-level men's basketball players. *J. Strength Cond. Res.* **2010**, *24*, 1346–1355. [CrossRef] [PubMed]
13. Ribeiro, B.G.; Mota, H.R.; Sampaio-Jorge, F.; Morales, A.P.; Leite, T.C. Correlation between body composition and the performance of vertical jumps in basketball players. *J. Exerc. Physiol. Online* **2015**, *18*, 69–79.
14. Sisodiya, A.; Yadaf, M. Relationship of Anthropometric Variables to Basketball Playing Ability. *J. Adv. Dev. Res.* **2010**, *1*, 191–194.
15. Sallet, P.; Perrier, D.; Ferret, J.; Vitelli, V.; Baverel, G. Physiological differences in professional basketball players as a function of playing position and level of play. *J. Sports Med. Phys. Fit.* **2005**, *45*, 291–294.

16. Vuckovic, I.; Mekic, M. Morphological characteristics of basketball players from playing position aspect. In *1st International Scientific Conference Exercise and Quality of Life Novi Sad*; Mikalacki, M., Ed.; University of Novi Sad: Novi Sad, Serbia, 2009; pp. 309–316.

17. Purenović-Ivanović, T.; Popović, R. Somatotype of top-level Serbian rhythmic gymnasts. *J. Hum. Kinet.* **2014**, *40*, 181–187. [CrossRef] [PubMed]

18. Popovic, S.; Akpinar, S.; Jaksic, D.; Matic, R.; Bjelica, D. Comparative study of anthropometric measurement and body composition between elite soccer and basketball players. *Int. J. Morphol.* **2013**, *31*, 461–467. [CrossRef]

19. Malina, R.M.; Kozieł, S.M. Validation of maturity offset in a longitudinal sample of polish boys. *J. Sports Sci.* **2013**, *32*, 424–437. [CrossRef] [PubMed]

20. Stewart, A.; Marfell-Jones, M.J.; Olds, T.; De Ridder, H. *International Standards for Anthropometric Assessment*; International Society for Advancement of Kinanthropometry: Lower Hutt, New Zealand, 2012.

21. Carter, J.L.; Heath, B.H. *Somatotyping: Development and Applications*; Cambridge University Press: Cambridge, UK, 1990.

22. Association, W.M. World Medical Association Declaration of Helsinki. Ethical principles for medical research involving human subjects. *Bull. World Health Organ.* **2001**, *79*, 373–374.

23. Hays, W.L. *Statistics Fort Worth*; Harcourt Brace College Publications: Fort Worth, TX, USA, 1994.

24. Cohen, L.; Manion, L.; Morrison, K. *Research Methods in Education*; Routledge: Abingdon, UK, 2013.

25. Grissom, R.J.; Kim, J.J. *Effect Sizes for Research: Univariate and Multivariate Applications*; Routledge: Abingdon, UK, 2012.

26. Jaszczanin, J.; Palonka, M.; Buryta, R.; Krupecki, K.; Cieszczyk, P. Cechy somatyczne koszykarzy ekipy narodowej polski i zespołów uczestniczących na igrzyskach olimpijskich Sydney. *Annales Universitatis Marie Curie-Składowska* **2004**, *59*, 356–360.

27. Moxley, J.H.; Towne, T.J. Predicting success in the national basketball association: Stability & potential. *Psychol. Sport Exerc.* **2015**, *16*, 128–136.

28. Litkowycz, R.; Zając, A.; Waśkiewicz, Z. Zmienność ontogenetyczna predyspozycji morfologiczno-strukturalnych koszykarzy na różnych etapach szkolenia sportowego. *Antropomotoryka* **2015**, *30*, 39–45.

29. Latin, R.W.; Berg, K.; Baechle, T. Physical and Performance Characteristics of NCAA Division I Male Basketball Players. *J. Strength Cond. Res.* **1994**, *8*, 214–218.

30. Delextrat, A.; Cohen, D. Physiological testing of basketball players: Toward a standard evaluation of anaerobic fitness. *J. Strength Cond. Res.* **2008**, *22*, 1066–1072. [CrossRef] [PubMed]

31. Martinez, P.Y.O.; López, J.A.H.; Meza, E.I.A.; Arráyales, M.E.M.; Sánchez, L.R. Somatotype Profile and Body Composition of Players from the Mexican Professional Basketball League. *Int. J. Morphol.* **2014**, *32*, 1032–1035. [CrossRef]

32. Rahmawati, N.T.; Budiharjo, S.; Ashizawa, K. Somatotypes of young male athletes and non-athlete students in Yogyakarta, Indonesia. *Anthr. Sci.* **2007**, *115*, 1–7. [CrossRef]

33. Handziska, E.; Handziski, Z.; Gjorgoski, I.; Dalip, M. Somatotype and stress hormone levels in young soccer players. *J. Sports Med. Phys. Fit.* **2015**, *55*, 1336–1342.

34. George, M.; Evangelos, T.; Alexandros, K.; Athanasios, L. The inside game in World Basketball. Comparison between European and NBA teams. *Int. J. Perform. Anal. Sport* **2009**, *9*, 157–164. [CrossRef]

35. Kułaga, Z.; Litwin, M.; Tkaczyk, M.; Palczewska, I.; Zajączkowska, M.; Zwolińska, D.; Krynicki, T.; Wasilewska, A.; Moczulska, A.; Morawiec-Knysak, A.; et al. Polish 2010 growth references for school-aged children and adolescents. *Eur. J. Pediatr.* **2011**, *170*, 599–609. [CrossRef] [PubMed]

36. Malina, R.M. Skeletal age and age verification in youth sport. *Sports Med.* **2011**, *41*, 925–947. [CrossRef] [PubMed]

37. Le Gall, F.; Carling, C.; Williams, M.; Reilly, T. Anthropometric and fitness characteristics of international, professional and amateur male graduate soccer players from an elite youth academy. *J. Sci. Med. Sport* **2010**, *13*, 90–95. [CrossRef] [PubMed]

38. Malina, R.M.; Eisenmann, J.C.; Cumming, S.P.; Ribeiro, B.; Aroso, J. Maturity-associated variation in the growth and functional capacities of youth football (soccer) players 13–15 years. *Eur. J. Appl. Physiol.* **2004**, *91*, 555–562. [CrossRef] [PubMed]

© 2018 by the authors. Licensee MDPI, Basel, Switzerland. This article is an open access article distributed under the terms and conditions of the Creative Commons Attribution (CC BY) license (http://creativecommons.org/licenses/by/4.0/).

sports

MDPI

Article

Relation between Motor and Cognitive Skills in Italian Basketball Players Aged between 7 and 10 Years Old

Francesca Policastro [1,*] [iD], Agostino Accardo [2], Roberto Marcovich [3], Giovanna Pelamatti [1] and Stefania Zoia [1]

1 Life Science Department, University of Trieste, 34100 Trieste, Italy; pelamatti@units.it (G.P.); stefania.zoia@asuits.sanita.fvg.it (S.Z.)
2 Architecture and Engineering Department, University of Trieste, 34100 Trieste, Italy; accardo@units.it
3 Medical Science Department, University of Trieste, 34100 Trieste, Italy; marcovic@units.it
* Correspondence: frapol7@hotmail.com

Received: 11 July 2018; Accepted: 13 August 2018; Published: 14 August 2018

Abstract: There is evidence supporting a correlation between motor, attention and working memory in children. This present study focuses on children aged between 7 and 10 years, who have been playing basketball in the last two years. The aim of this study is to verify the correlation between cognitive and motor abilities and to understand the importance of this correlation in basketball practice. A total of 75 children who were 7.2–10.99 years old were assessed in terms of their attention, motor manual sequences and visuo-spatial working memory. A regression analysis was provided. In this sample, the motor abilities of children were found to be correlated with attention (denomination task, $R^2 = 0.07$), visuo-spatial working memory ($R^2 = 0.06$) and motor manual sequencing (aiming and catching task, $R^2 = 0.05$; and manual dexterity task, $R^2 = 0.10$). These correlations justify the suggestion to introduce deeper cognitive involvement during basketball training. The development of executive functions could have an important impact on basketball practice and the introduction of attention and memory tasks could help coaches to obtain optimal improvement in performance during the training sessions.

Keywords: basketball; Movement Assessment Battery for Children-2; attention; visuo-spatial working memory; motor manual sequences

---

1. Introduction

Basketball is regarded as one of the most popular sports worldwide. A considerable number of players start practicing basketball as early as 5–6 years of age. In the United States of America (USA), the National Basketball Association and the USA Basketball league created the "Youth Basketball Guidelines" [1], which aimed to promote the physical health of players; to develop age- and stage-appropriate skills; and to foster the development of peer relationships, self-esteem and leadership qualities. These guidelines provide age-appropriate standards that follow the maturation of children. The guidelines focus on the game structure (i.e., game length, timeouts), the tactics (i.e., how to set the defence) and the rules (i.e., how to manage substitutions). In Italy, the basketball association (Federazione Italiana Pallacanestro) involves more than 300,000 coaches, players and young players [2]. In fact, basketball is the second most popular sport after soccer.

Italian basketball coaches for young players complete two-year vocational training before transitioning to practice. During this training period, the focus is largely placed on the physical, social and emotional development of the participating children [3]. Their training needs to include adequate consideration of the aerobic resistance, motor abilities and socio-emotional development of the children [4]. There have been suggestions that training should also engage cognition. For instance, the coaches should incorporate problem-solving games in order to try to foster imaginative processes [5] and develop timing and spacing abilities in their players. In any case, there is a lack of knowledge and specific information that is required for encouraging the cognitive development of these young children during basketball training. Considering the importance of cognitive abilities in motor learning, this lack of knowledge on cognitive skills could have an impact on training. Newell's Theory of Constraints [6] provides a complete framework for the correlation between cognitive and motor abilities. The type of task, the environment and the capabilities of the individual player influence the motor performance. For instance, the quality of a jump depends on the environmental constraints (i.e., the surface type, the environmental stability), on the task demand (i.e., to jump beyond an object, one/two legs) and on personal characteristics (i.e., strength, cognition, sensitivity). Through this model, Newell demonstrated the reciprocal integration that exists between the dynamic motor and cognitive systems. There is evidence that supports this integration, which was obtained from both healthy [7,8] and clinical samples [9,10]. Focusing on motor control, Coker [11] provided an example of this integration in basketball practice. In fact, motion and cognition can collaborate as an integrated system to provide the motor control of the gestures. Motor control focuses on the neural, physical and behavioural aspects that are necessary to produce the correct movements [12]. In basketball practice, these systems are required to refine and re-learn motor-skills, such as intercepting a ball at the correct time or improving the landing biomechanics to prevent injuries. Prerequisite abilities, such as control precision, multi-limb coordination, rate control, aiming and catching, timing control and dynamic flexibilities, are necessary for learning basketball. In this sense, cognitive and motor systems can be integrated to guarantee the best possible motor performance. Evidence has demonstrated that physical activity has an important impact on the development of the executive functions in children [13,14].

In this study, we hypothesised the opposite as we suggested that the introduction of a cognitive aspect would have an impact on basketball practice of children between 7 and 11 years old. In fact, this study focuses on the importance of a keen consideration of the cognitive aspects during basketball learning and practice. To introduce this type of approach, it is essential to analyse the relations between motor and cognitive skills of this particular sample. The preliminary aim of this study is to verify how the young basketball players normally develop motor and cognitive abilities, especially considering attention and memory functions. The purpose is not to compare basketball players to other athletes but to assess them with respect to the general Italian population. Furthermore, we wanted to understand if and how attention and memory abilities could be stimulated during basketball practice. Considering actual basketball practice, the main aim of this study is to propose the involvement of some specific executive functions tasks during basketball practice. Using the following results, we will be able to acquire a deeper understanding of the development of these young basketball players in order to provide some suggestions to coaches about the relevance of cognitive development in basketball practice.

2. Materials and Methods

2.1. Procedure

First, the project was sent to and accepted by the Ethic Committee of the University of Trieste. Following ethical approval, four basketball clubs in the region of Trieste (Italy) were invited to participate in this project. The data were collected and shared with the clubs to assess the development of the young players. A total of 116 parents gave their written informed consent to let their children participate in this study. These parents also filled an anamnestic questionnaire about the health status and the development of the child. The participants were assessed during training in an ecological situation but in a quiet and reserved part of the gym. Testing was conducted in two 30–45 min sessions for each child. For each subject, the sessions were both completed in two months, which were always conducted by the same operator. This study started on January 2016 and ended on February 2017.

Testing was randomised in order to eliminate order-related biases. The participants showed interest in the project and were motivated.

2.2. Participants

A total of 116 participants (79 boys and 37 girls) were recruited. The mean age was 9.23 years (SD = 1.07). The participation during this study depended on the participation of the children in basketball training. For this reason, the final sample included only 75 children (53 boys and 22 girls) who were aged between 7.23 and 10.99 years (mean = 9.36, SD = 0.98).

The inclusion criteria were being children that are aged 7–10.99 years old and playing basketball for at least the last two years. Furthermore, the participants must have practiced basketball two or three times per week at a competitive level. The basketball activity is not linked to the school program and there were no academic penalties if the children did not participate.

Using the parents' questionnaire, one participant with Attention Deficit Hyperactivity Disorder (ADHD) was excluded from the study.

2.3. Measures

To assess motor and cognitive skills, the following neuropsychological tests were used. Each test provides specific instructions to be administered in the correct manner, which allowed us to create trials with different controls.

In order to provide a complete assessment of cognitive and motor skills, the proposed tasks are at different difficulty levels, which are useful in characterising healthy and sporty children.

- Movement Assessment Battery for Children—2 (MABC-2) [15].

The MABC-2 is a standardised test that is used to assess the motor skills of children with movement difficulties in the following domains: Manual dexterity (MD), aiming and catching (AC) and balance (Bal). Cluster and total standard scores for Italian children are provided [16], with higher scores demonstrating better performance. A total test score at or below the 5th percentile indicates significant movement difficulty, while a score between the 5th and 15th percentile indicates that a child is "at risk".

In these tests, the items and the scores were compared to the normative data of the uploaded version of the Italian MABC-2.

- Attention, Inhibition and Switching Assessment from the Neuropsychological Assessment—2 (NEPSY-II) [17].

This test is included in the "Attention and Executive Functioning" domain. It requires each participant to look at a series of black and white shapes or arrows and name the shape, the direction or an alternate response depending on the colour of the shape or arrow. The subtests of denomination, inhibition and switching provide information about the accuracy of the sample in terms of timing and errors made when completing the tests. A comparison between the plots that show the relationship between time and errors is not possible as many scores were 0.

In these tests, the items and the scores were compared to the normative data of the uploaded version of the Italian NEPSY-II.

- Manual Motor Sequences Assessment (MMSA) from NEPSY-II

This test is included in the "Sensorimotor" domain. "Manual Motor Sequences" requires each participant to imitate and repeat a series of hand movements performed by the examiner. This counts the number of the manual sequences that participants can replicate.

In this test, the items and the scores were compared to the normative data of the uploaded version of the Italian NEPSY-II.

- Corsi's Test—Sequential Spatial Task [18].

This test assesses visuo-spatial short-term working memory. Each participant should imitate the examiner who taps a sequence of up to nine identical spatially separated blocks. Each participant should also be able to repeat this tapping sequence backwards.

2.4. Statistical Analysis

We conducted statistical analysis using Matrix Laboratory (MatLab). First, the sample was compared to the normative data given by the tests in a qualitative and in a quantitative way whenever possible. For the quantitative assessment, we used the unpaired *t*-test (*p*-value < 0.05) to compare the observed data to the full data set of normative values. For MABC-2 and its subtests, the scores are given in percentiles [15]. Thus, there were no comparisons made using *p*-values but qualitative considerations were undertaken, which were related to the incidence of motor deficits. For the three attention tasks, qualitative suggestions are provided due to the structure of the given standardised data, which consider the accuracy in terms of time and error relations [17]. For MMSA and for the Corsi's Test, it was possible to obtain a quantitative comparison [17,18].

Second, the scores obtained by the participants were used to provide a regression analysis of the correlations between MABC-2 and attention, visuo-spatial working memory (Corsi's Test) and motor manual sequencing (MMSA).

All the chosen tests are internationally validated and used. Furthermore, they have also been adapted and standardised for the Italian population. The differences found between the countries highlight the importance in considering the cultural variables. For this reason, the standard population given by the tests allowed comparison with the subjects of the study. Comparing the sample to a standard expected normative Italian population could partially reduce the limitations of not having a control group.

3. Results

3.1. Data Analysis and Description

Table 1 shows the means, the standard deviations, the ranges and the expected values for the considered variables. All the scores are age-standardised.

Table 1. Means, SDs and ranges of score for the study variables.

Test	Mean	SD	Range
MABC-2 ^,*	73	20.9	5–98
MD ^,*	57	25.8	1–95
AC ^,*	76	18.3	9–99.9
Bal ^,*	68	18.0	16–95
Denomination ^	11.4	2.8	4–17
Inhibition ^	10.4	2.4	6–14
Switching ^	10.8	2.7	6–16
MMSA ^	10.8	2.3	2–14
Corsi's Test ^	4.4	1.1	2–7

* Score in percentiles, ^ Age-standardised score.

The sample of children in this present study had a mean rank in the 73rd percentile in the MABC-2 test. Two children scored lower than the critical score. Both children had a very low score in the MD cluster (1st percentile). MD is also the cluster with the lower mean-score.

In NEPSY-II attention test, it is important to highlight that the majority of the observed sample had higher ranks in each subtest compared to the standard population (59% in denomination, 60% in inhibition and 64% in switching).

In NEPSY-II MMSA test, the sample demonstrates statistically significant higher scores compared to the standard population provided by the test, with a p-value of 0.017 (standard population: mean-score = 10, SD = 3).

In Corsi's test, the mean-score of the sample was 4.4 (SD = 1.15), with this test having a normal range of 2–7 points. The standardised scores depend on the age of the child and change for every year. A brief stratification of ages and the unpaired t-test scores (see Table 2) demonstrated that there were no significant differences between the sample and the normal population. Nine children scored lower than the critical score for their age.

Table 2. Stratification for ages of Corsi's Test.

Age	Mean	SD	Expected Mean	t-Value	p-Value	n
10 years	5.00	1.03	4.37	0.85	0.41	20
9 years	3.96	0.95	4.35	0.57	0.57	24
8 years	4.44	1.29	4.22	0.24	0.81	25
7 years	3.83	0.75	4.03	0.35	0.74	6

Note: No significant p-values.

3.2. Regression Analysis

Table 3 contains all pairs of variables that had a significant correlation (p-value < 0.05). The other correlations were not reported because they do not have an impact on the aim of this study.

First, Table 3 summarises the regression results for the MMSA by showing just the statistically significant correlations. It can be observed that MMSA is moderately correlated with the total scores of MABC-2, MD and AC.

Second, this table summarises the significant regression results for the attention task performances. The denomination subtest demonstrates a strong correlation with motor skills, while the inhibition subtest is related to MMSA. The switching test is the only subtest of attention that is not related to other variables.

Finally, the table summarises the significant regression results from Corsi's Test. There still exists a correlation with the motor skills. It is important to emphasise that the two children who scored very low in the MABC-2 test also had low scores on Corsi's test.

Table 3. Significant results from regression analysis.

DV	IV	Bs	Bi	SE	Beta	R^2	p-Value
MMSA	MABC-2	3.35	36.85	21.1	0.34	0.12	0.003
MMSA	MD	3.76	14.70	26.1	0.31	0.10	0.006
MMSA	AC	1.97	57.93	20.3	0.22	0.05	0.050
Denomination	MABC-2	2.08	49.58	21.7	0.27	0.07	0.021
Inhibition	MMSA	0.33	7.35	2.2	0.35	0.12	0.002
Corsi's Test	Inhibition	0.48	8.27	2.4	0.23	0.05	0.046
Corsi's Test	MABC-2	4.58	53.07	21.9	0.24	0.06	0.042

MMSA = Motor Manual Sequences Assessment; MABC-2 = Movement Assessment Battery for Children-2; MD = Manual Dexterity; AC = Aiming and Catching; DV = dependent variable; IV = independent variable; Bs = Beta slope; Bi = Beta intercept; SE = standard error for Beta; Beta = standardised Beta; and R^2 = correlation coefficient.

4. Discussion

The first aim of this study was to identify the correlations between executive functions and motor abilities in this specific sample. The correlations would support our suggestion that coaches should consider these aspects in training. The description and the analysis of the data demonstrate some positive correlations between motor and cognitive skills in this basketball sample. According to the previous evidence [7,19], the cognitive aspects involved in the development of the motor skills are:

Attention [8] (in terms of inhibition and denomination), visuo-spatial working memory [9,10] and sensorimotor ability to imitate motor manual sequences. The results from the present study were consistent with the research of Roebers [20] and Mandlich [21], who demonstrated the involvement of attention in the development of motor skills. We found that performance on the denomination task is correlated with the MABC-2 score ($R^2 = 0.07$). For instance, the basketball players frequently utilise their denomination skills when they focus on the number and the position of the player that they are defending. Attention is a necessary executive function for developing correct movement strategies and inhibiting unwanted movement. We found a correlation between the inhibition subtest and the motor manual sequencing ($R^2 = 0.12$). Inhibition and motor manual sequencing abilities are involved in the sensorimotor aspect of movement [12], which occurs when the players do not pass the ball to a companion, when an opponent quickly appears or when they stop moving in order to avoid penalties. The correlations of MMSA with AC ($R^2 = 0.05$) and MD ($R^2 = 0.10$) explain the importance of the motor manual sequencing in implementing a motor action plan. We evaluated these outcomes as being necessary for young basketball players due to the important participation of upper limbs in movement control during basketball practice. For instance, players utilise their ability to imitate motor manual sequences every time they are learning a manual task by imitating the coach, such as during the ball-handling practice. Consistent with the research of Alloway and Temple [9], the present study also demonstrates the significant role of visuo-spatial working memory through the correlation found between the Corsi's test and the MABC-2 scores ($R^2 = 0.06$). Basketball practice is based on visuo-spatial structures, which are used to create and improve complex motor tasks, thus creating "automatic" movements to play. For instance, players use their visuo-spatial working-memory when they learn how to dribble around opponents and when they recall this and other complex automated motor plans during the game. The previous examples highlight how motor and cognitive systems are integrated to guarantee complex controlled motor performance [6]. These capabilities are necessary in the development of basketball practice because they also contribute to the sensorimotor aspect of the movement and permit the development of rapid motor responses during the game [10,21]. The physiological explication of these correlations comes from the shared neural mechanism, especially of cerebellum processes [19,22,23].

The main aim of the study was to consider the introduction of specific cognitive tasks during basketball practice in order to optimise and enrich training potential. Furthermore, children aged between 7 and 11 years old, which formed the sample in this present study, are experiencing important cognitive development as their consciousness of their own cognitive abilities increased in these years. Starting from this developmental period, children are able to identify attention and memory tasks. They can discriminate the meaning and the functioning of the different types of attention and memory [24]. The introduction of cognitive-aware tasks during training could facilitate motor learning itself. Therefore, considering the cognitive dynamic system in basketball learning could provide players with a deeper understanding of their own movements.

Finally, the results from the present study suggest that coaches should train attention, visuo-spatial working memory and motor manual sequencing ability of their young players. A targeted proposal for young basketball players could help sport-specific learning and could optimise and enrich the players' abilities.

As mentioned above, we do not want to emphasise that basketball practice is the cause of the correlations found in this study. We want to instead focus on the fact that the presence of these correlations (specific for this basketball sample) could have an impact on basketball learning in young players.

In the future, it could be useful to increase the sample size to confirm the present correlations and to verify whether there is a corresponding increase in their magnitude. It could also be interesting to create and validate a practical training proposal, which would involve training of the executive functions.

The strengths of this study include the practical proposal of cognitive training of children in basketball and the basis of this proposal on the significant statistical correlations found between

Sports **2018**, *6*, 80

motor and cognitive skills. The limitations of this present study include the moderate strength of the correlations, which was possibly caused by the small sample size, and the inability to quantitatively analyse all data.

Author Contributions: Conceptualization, F.P. and S.Z.; Methodology, F.P.; Software, A.A.; Validation, S.Z. and G.P.; Formal Analysis, F.P. and S.Z.; Investigation, F.P., S.Z. and A.A.; Resources, R.M.; Data Curation, F.P.; Writing-Original Draft Preparation, F.P.; Writing-Review & Editing, F.P.; Visualization, S.Z. and A.A.; Supervision, G.P and S.Z.

Funding: This research did not receive any specific grant from funding agencies in the public, commercial, or not-for-profit sectors.

Acknowledgments: We are very grateful to the basketball clubs, the parents and the children who participated in this study.

Conflicts of Interest: The authors declare no conflicts of interest.

References

1. National Basketball Association. USA Basketball Youth Basketball Guidelines. 2017. Available online: https://youthguidelines.nba.com (accessed on 2 July 2018).
2. Federazione Italiana Pallacanestro. Statistiche Tesserati a Livello Nazionale e Regionale. 2012. Available online: http://www.fip.it/trentinoaltoadige/DocumentoDett.asp?IDDocumento=51818 (accessed on 2 July 2018).
3. Mondoni, M.; Avanzino, R. *Manuale di pallacanestro. Tecnica e didattica*; Vita e pensiero, Università di Cattolica: Milan, Italy, 2014; ISBN 10:8834324994.
4. Cremonini, M.; Pellegrino, F.M. *Minibasket. L'emozione, la Scoperta, il Gioco*; Federazione Italiana Pallacanestro: Milan, Italy, 2010.
5. Cremonini, M. Le Favole Nel Minibasket. 2002. Available online: http://www.basketservice.it/AREA%20TECNICA/MINIBASKET/MBD070.pdf (accessed on 2 July 2018).
6. Newell, K.M. Coordination, control and skill. *Adv. Psychol.* **1985**, *27*, 295–317.
7. Piek, J.P.; Dawson, L.; Smith, L.M.; Gasson, N. The role of early fine and gross motor development on later motor and cognitive ability. *Hum. Mov. Sci.* **2008**, *27*, 668–681. [CrossRef] [PubMed]
8. Niederer, I.; Kriemler, S.; Gut, J.; Hartmann, T.; Schindler, C.; Barral, J.; Puder, J.J. Relationship of aerobic fitness and motor skills with memory and attention in preschoolers (Ballabeina): A cross-sectional and longitudinal study. *BMC Pediatr.* **2011**, *11*, 34. [CrossRef] [PubMed]
9. Alloway, T.P.; Temple, K.J. A comparison of working memory skills and learning in children with developmental coordiantion disorder and moderate learning difficulties. *Appl. Cogn. Psychol.* **2007**, *21*, 473–487. [CrossRef]
10. Piek, J.P.; Dyck, M.J.; Francis, M.; Conwell, A. Working memory, processing speed, and set-shifting in children with developmental coordination disorder and attention-deficit–hyperactivity disorder. *Dev. Med. Child Neurol.* **2007**, *49*, 678–683. [CrossRef] [PubMed]
11. Coker, C.A. *Motor Learning and Control for Practitioners*; Routledge: Abingdon-on-Thames, UK, 2017.
12. Rosenbaum, D.A. *Human Motor Control*; Academic Press: Cambridge, MA, USA, 2009.
13. Diamond, A.; Lee, K. Interventions shown to aid executive function development in children 4 to 12 years old. *Science* **2011**, *333*, 959–964. [CrossRef] [PubMed]
14. Benzing, V.; Schmidt, M.; Jäger, K.; Egger, F.; Conzelmann, A.; Roebers, C.M. A classroom intervention to improve executive functions in late primary school children: Too 'old' for improvements? *Br. J. Educ. Psychol.* **2018**. [CrossRef] [PubMed]
15. Henderson, S.E.; Sugden, D.A.; Barnett, A.L. *Movement Assessment Battery for Children-2*; Harcourt Assessment, Psychological Corporation: London, UK, 2007.
16. Biancotto, M.; Guicciardi, M.; Pelamatti, G.M.; Santamaria, T.; Zoia, S. *Movement Assessment Battery for Children*, 2nd ed.; Giunti O.S. Psychometrics: Firenze, Italy, 2016; ISBN 978-88-09-9942-B.
17. Korkman, M.; Kirk, U.; Kemp, S.L. *NEPSY: A Developmental Neuropsycological Assessment*, 2nd ed.; The Psychological Corporation: San Antonio, TX, USA, 2007.
18. Mammarella, I.C.; Toso, C.; Pazzaglia, F.; Cornoldi, C. *BVS-Corsi. Batteria per la Valutazione Della Memoria Visiva e Spaziale*; Con CD-ROM; Edizioni Erickson: Trento, Italia, 2008.

19. Rigoli, D.; Piek, J.P.; Kane, R.; Oosterlaan, J. An examination of the relationship between motor coordination and executive functions in adolescents. *Dev. Med. Child Neurol.* **2012**, *54*, 1025–1031. [CrossRef] [PubMed]

20. Roebers, C.M.; Kauer, M. Motor and cognitive control in a normative sample of 7-years-olds. *Dev. Sci.* **2009**, *12*, 175–181. [CrossRef] [PubMed]

21. Mandich, A.; Buckolz, E.; Polatajko, H. On the ability of children with developmental coordination disorder (DCD) to inhibit response initiation: The Simon effect. *Brain Cogn.* **2002**, *50*, 150–162. [CrossRef]

22. Diamond, A. Close interrelation of motor development and cognitive development and of the cerebellum and prefrontal cortex. *Child Dev.* **2000**, *71*, 44–56. [CrossRef] [PubMed]

23. Best, J.R. Effects of physical activity on children's executive function: Contributions of experimental research on aerobic exercise. *Dev. Rev.* **2010**, *30*, 331–351. [CrossRef] [PubMed]

24. Blair, C.; Zelazo, P.D.; Greenberg, M.T. *Measurement of Executive Function in Early Childhood: A Special Issue of Developmental Neuropsychology*; Psychology Press: East Sussex, UK, 2016.

© 2018 by the authors. Licensee MDPI, Basel, Switzerland. This article is an open access article distributed under the terms and conditions of the Creative Commons Attribution (CC BY) license (http://creativecommons.org/licenses/by/4.0/).

sports

MDPI

Article

Regional Differences in Women's Basketball: A Comparison among Continental Championships

Haruhiko Madarame [ID]

Department of Sports and Fitness, Shigakkan University, Nakoyama 55, Yokonemachi, Obu, Aichi 474-8651, Japan; madarame-tky@umin.ac.jp; Tel.: +81-562-46-1291

Received: 27 June 2018; Accepted: 18 July 2018; Published: 20 July 2018

Abstract: The aims of this study were (i) to compare basketball game-related statistics in women by region (Africa, America, Asia, Europe), and (ii) to identify characteristics that discriminate performances for each region. A total of 134 games from each continental championship held in 2017 were analyzed. A one-way ANOVA followed by a Bonferroni-adjusted pairwise comparison was performed to evaluate differences in each variable between the continents. A discriminant analysis was performed to identify game-related statistics that discriminate among the continents. The Asian and European championships overall showed similar performance profiles: Low numbers of possessions and turnovers, and high numbers of successful field goals and assists. However, the European championship was more closely contested than the Asian championship. The African championship was characterized by high numbers of possessions, free throws, and turnovers. The homogeneity of the American championship was low, and some of the cases have similarities with the African championship, whereas other cases have similarities with the European championship. On average, the American championship was characterized by low numbers of successful field goals and assists, and high numbers of steals and turnovers. It is suggested that women's basketball games are played in a different manner in each region of the world.

Keywords: basketball; game-related statistics; performance analysis; team sports

1. Introduction

Basketball is one of the most popular sports in the world. As of 2018, the International Basketball Federation (FIBA) has 213 national member federations, and the FIBA estimates that there are 450 million players worldwide [1]. National teams compete in international competitions, such as the Olympic Basketball Tournament, the FIBA World Cup, and the FIBA Continental Cups. International competitions in basketball are governed by the FIBA, so that official games are played by the same rules with the same equipment anywhere in the world. However, regional differences in performance profiles have been reported in recent studies [2,3]. Ibáñez et al. [2] compared game-related statistics among continental championships for men held in 2015, and reported that each continent has a specific performance profile, which can be summarized as follows: Africa, high numbers of free throws, rebounds, steals, and fouls; America, a high number of field goal attempts; Asia, a high number of possessions and a low number of assists; and Europe, a low number of possessions and a high number of assists. These findings indicate that basketball games are played in a different manner in each continent.

From a practical perspective, the knowledge about regional differences in performance profiles would be useful for players and coaches of national teams preparing for international competitions. However, although international competitions are held not only for men, but also for women, previous studies on regional differences in basketball [2,3] have analyzed only men's competitions. Sex differences in performance profiles have been reported in previous studies [4–6]. For example,

Sampaio et al. [4] analyzed the world championships for both men and women held in 2002 and reported that men's teams were discriminated from women's teams by a higher percentage of blocks and a lower percentage of steals, suggesting that anthropometric differences between men and women might be attributable to the difference in performance profiles. Differences between men and women can also be found in the latest FIBA World Rankings [7,8]. In the men's ranking, updated as of 28 February 2018 [7], the 10th-ranked Australia is the only country that ranks in the top 20 and belongs neither to America nor to Europe. In the women's ranking, updated as of 27 August 2017 [8], however, the top 20 includes four countries from Asia (Australia, China, Japan, Korea) and two countries from Africa (Senegal, Angola) (Note that, although Australia belongs to FIBA Oceania, Oceanian championships have been merged with Asian championships since 2017, and Oceanian countries have been categorized into Asia in the FIBA World Rankings). Considering these facts, it is possible to assume that regional differences in performance profiles differ between men and women. If regional differences among continental championships for women are dissimilar to those for men, the previous findings on men [2,3] cannot be applied to women. Therefore, identifying regional differences in women's basketball would be of help for players and coaches of women's national teams to prepare for international competitions.

The number of studies on game-related statistics in women's basketball has been increasing in recent years [9–15]. Game-related statistics have been analyzed to identify the relationship between performance indicators and match outcomes in international [9–11] and domestic [12–14] tournaments, and to identify performance indicators that discriminate starters from nonstarters in a professional league [15]. Although one study has investigated game-related statistics that discriminate winners from losers in both Asian and European women's championships held in 2011, 2013, and 2015 [11], no studies have investigated regional differences in women's basketball among four continental championships (Africa, America, Asia, Europe). Therefore, the aims of this study were (i) to compare basketball game-related statistics in women by region (Africa, America, Asia, Europe), and (ii) to identify characteristics that discriminate performances for each region.

2. Materials and Methods

Box scores of all 134 games in four continental championships for women held in 2017 (Table 1) were gathered from the official website of FIBA.

Table 1. Sample characteristics.

Region	Teams	Games	Cases
Africa	12 (Angola, Central African Republic, Cameroon, Cote d'Ivoire, Dem. Rep. of Congo, Egypt, Guinea, Mali, Mozambique, Nigeria, Senegal, Tunisia)	46	92
America	10 (Argentina, Brazil, Canada, Colombia, Cuba, Mexico, Paraguay, Puerto Rico, Venezuela, Virgin Islands)	24	48
Asia	8 (Australia, China, Chinese Taipei, D.P.R. of Korea, Japan, Korea, New Zealand, Philippines)	24	48
Europe	16 (Belarus, Belgium, Czech Republic, France, Greece, Hungary, Italy, Latvia, Montenegro, Russia, Serbia, Slovak Republic, Slovenia, Spain, Turkey, Ukraine)	40	80
Total	46	134	268

Data reliability of the box scores was not assessed in this study. However, official box scores are treated as reliable in basketball studies [16,17], because the recording process is executed according to the regulations established by FIBA [18], and a high level of inter-rater reliability (kappa coefficient above 0.89) has been repeatedly confirmed [2,19–21]. Game-related statistics of each game were analyzed separately for the winning and losing teams, so that 268 cases were analyzed in total. The analyzed game-related statistics were as follows: 2- and 3-point field goals (successful and unsuccessful), free throws (successful and unsuccessful), defensive and offensive rebounds, assists, steals, turnovers, blocks, and fouls committed. Definitions of the statistics [18] are shown in Table 2.

To eliminate the effect of game rhythm, the variables were normalized to 100 game ball possessions [22]. Game ball possessions were calculated as an average of team ball possessions (TBP) of both teams [23]. TBP was calculated from field goal attempts (FGA), offensive rebounds (ORB), turnovers (TO), and free throw attempts (FTA) using the following equation [23]:

$$TBP = FGA - ORB + TO + 0.4 \times FTA \qquad (1)$$

Statistical analyses were performed with R version 3.5.0 for Windows [24]. Statistical significance was set at $p \leq 0.05$ unless otherwise stated. A one-way analysis of variance followed by a Bonferroni-adjusted pairwise comparison was performed to evaluate differences in each variable between the continents. Cohen's d was calculated as an effect size and interpreted as follows: $d = 0.20$ to 0.49, small effect; $d = 0.50$ to 0.79, medium effect; $d \geq 0.80$, large effect [25]. To identify game-related statistics that discriminate between the continents, a discriminant analysis was performed using R code 'candis' and 'geneig', which have been used in previous studies [3,6,11,26]. An absolute value of a structural coefficient (SC) greater than or equal to 0.30 was considered relevant for the discrimination between the continents [2,3].

3. Results

Significant *F*-values were obtained for point difference, team ball possessions, successful 2- and 3-point field goals, successful and unsuccessful free throws, assists, steals, turnovers, and fouls committed (Table 3). Large effect size differences between each continent were observed for point difference (Africa vs. Europe, $d = 0.85$; Asia vs. Europe, $d = 0.97$), team ball possessions (Africa vs. Europe, $d = 1.08$; America vs. Europe, $d = 1.45$), unsuccessful free throws (Africa vs. Asia, $d = 1.16$; Africa vs. Europe, $d = 0.82$), assists (America vs. Asia, $d = 1.03$; America vs. Europe, $d = 0.81$), and fouls committed (America vs. Europe, $d = 0.93$; Asia vs. Europe, $d = 1.05$) (Table 3).

Table 2. Definitions of game-related statistics according to International Basketball Federation (FIBA) statisticians' manual.

Statistics	Definitions
2-point field goals	A 2-point field goal attempt is charged to a player any time he shoots, throws, or tips a live ball at his opponent's basket in an attempt to score a goal from the 2-point field goal area.
3-point field goals	A 3-point field goal attempt is charged to a player any time he shoots, throws, or tips a live ball at his opponent's basket in an attempt to score a goal from the 3-point field goal area.
Free throws	A free throw is an opportunity given to a player to score one point, uncontested, from a position behind the free-throw line and inside the semi-circle.
Rebounds	A rebound is the controlled recovery of a live ball by a player or a team being entitled to the ball for a throw-in after a missed field goal attempt or last free throw attempt.
Assists	An assist is a pass that leads directly to a team-mate scoring. Scoring includes free throws. If the player who receives the pass is fouled in the act of shooting and makes at least one free throw, an assist is awarded in the same way as for a field goal made.
Steals	A steal is charged to a defensive player when his action causes a turnover by an opponent. A steal must always include touching the ball, but does not necessarily have to be controlled.
Turnovers	A turnover is a mistake by an offensive player or team that results in the defensive team gaining possession of the ball.
Blocks	A blocked shot is charged to a player any time he appreciably makes contact with the ball to alter the flight of a field goal attempt and the shot is missed.
Fouls committed	A foul is an infraction of the rules concerning illegal personal contact with an opponent and/or unsportsmanlike behavior.

Table 3. Game-related statistics of each continental championship with results of the ANOVA and post hoc comparisons.

Statistics	AF Mean	AF SD	AM Mean	AM SD	AS Mean	AS SD	EU Mean	EU SD	ANOVA F	ANOVA P	AF-AM P	AF-AM d	AF-AS P	AF-AS d	AF-EU P	AF-EU d	AM-AS P	AM-AS d	AM-EU P	AM-EU d	AS-EU P	AS-EU d
PTS	66.5	19.2	63.6	12.2	70.2	17.8	65.1	10.4	1.66	0.18	1.00	0.17	1.00	0.20	1.00	0.09	0.23	0.43	1.00	0.13	0.43	0.38
PD	25.3	19.4	16.0	11.9	24.8	17.9	12.3	8.4	13.07	**<0.01**	**<0.01**	0.54	1.00	0.03	**<0.01**	0.85	**0.03**	0.57	1.00	0.38	**<0.01**	**0.97**
TBP	78.2	7.7	78.1	5.4	75.1	6.3	71.5	3.9	20.48	**<0.01**	1.00	0.02	**0.03**	0.43	**<0.01**	**1.08**	0.11	0.50	**<0.01**	**1.45**	**<0.01**	0.73
S2P	23.9	8.2	22.9	6.2	28.0	9.3	26.3	5.8	5.32	**<0.01**	1.00	0.14	**0.01**	0.48	0.23	0.33	**<0.01**	0.65	0.07	0.57	1.00	0.23
U2P	35.2	7.7	34.8	9.6	34.6	8.6	35.3	7.9	0.09	0.96	1.00	0.05	1.00	0.07	1.00	0.01	1.00	0.02	1.00	0.06	1.00	0.08
S3P	6.2	3.7	6.7	4.0	8.0	4.1	7.5	3.4	3.14	**0.03**	1.00	0.12	**0.05**	0.46	0.13	0.38	0.53	0.32	1.00	0.24	1.00	0.12
U3P	16.8	6.6	18.0	5.6	17.5	7.4	17.2	5.6	0.42	0.74	1.00	0.19	1.00	0.11	1.00	0.07	1.00	0.07	1.00	0.14	1.00	0.04
SFT	18.5	7.2	15.8	5.4	13.7	5.9	15.9	6.8	6.06	**<0.01**	0.15	0.40	**<0.01**	0.70	0.07	0.36	0.67	0.38	1.00	0.02	0.38	0.35
UFT	10.1	4.8	6.8	3.1	5.1	3.1	6.5	3.9	21.09	**<0.01**	**<0.01**	0.78	**<0.01**	**1.16**	**<0.01**	**0.82**	0.26	0.54	1.00	0.08	0.37	0.38
DRB	36.7	7.6	39.3	8.3	36.4	8.5	37.9	5.9	1.68	0.17	0.32	0.33	1.00	0.04	1.00	0.18	0.37	0.34	1.00	0.20	1.00	0.22
ORB	17.8	7.5	15.8	6.5	15.9	7.3	16.2	7.1	1.20	0.31	0.78	0.27	0.94	0.24	0.98	0.21	1.00	0.02	1.00	0.06	1.00	0.04
AST	18.9	7.9	16.8	6.0	25.3	10.0	21.6	5.8	12.42	**<0.01**	0.69	0.29	**<0.01**	0.74	0.13	0.38	**<0.01**	**1.03**	**<0.01**	**0.81**	**0.04**	0.49
STL	11.7	5.1	12.0	5.2	9.6	5.0	10.8	4.3	2.65	**0.05**	1.00	0.06	0.10	0.42	1.00	0.19	0.10	0.47	1.00	0.26	1.00	0.26
TO	24.2	7.2	24.5	6.2	20.3	4.8	20.9	5.2	7.88	**<0.01**	1.00	0.04	**<0.01**	0.60	**<0.01**	0.51	**<0.01**	0.75	**<0.01**	0.63	1.00	0.12
BLK	2.9	2.7	3.2	2.5	3.5	2.5	3.6	2.6	1.19	0.31	1.00	0.10	1.00	0.23	0.50	0.26	1.00	0.13	1.00	0.16	1.00	0.03
FC	24.2	6.2	23.3	5.4	22.7	5.1	28.3	5.4	13.81	**<0.01**	1.00	0.16	0.72	0.27	**<0.01**	0.69	1.00	0.11	**<0.01**	**0.93**	**<0.01**	**1.05**

PTS, points scored; PD, point difference; TBP, team ball possessions; S2P, successful 2-point field goals; U2P, unsuccessful 2-point field goals; S3P, successful 3-point field goals; U3P, unsuccessful 3-point field goals; SFT, successful free throws; UFT, unsuccessful free throws; DRB, defensive rebounds; ORB, offensive rebounds; AST, assists; STL, steals; TO, turnovers; BLK, blocks; FC, fouls committed; $p \leq 0.05$ and $d \geq 0.80$ are shown in bold.

Classification results of the discriminant analysis are presented in Table 4. The total correct classification rate was 63.1%. Three significant functions were obtained from the discriminant analysis (Table 5). The territorial map of discriminant functions 1 and 2 is shown in Figure 1. The African and American championships were discriminated from the Asian and European championships by team ball possessions, unsuccessful free throws, assists, and turnovers (Function 1). The Asian championship was discriminated from the European championship by team ball possessions, assists, and fouls committed (Function 2). The African championship was discriminated from the American championship by successful free throws, unsuccessful free throws, and fouls committed (Function 3).

Table 4. Classification results of discriminant analysis.

Calculation	Region	Predicted				Total
		AF	AM	AS	EU	
Count	AF	**62**	10	8	12	92
	AM	14	**21**	2	11	48
	AS	6	1	**25**	16	48
	EU	10	3	6	**61**	80
Percentage	AF	**67.4**	10.9	8.7	13.0	100
	AM	29.2	**43.8**	4.2	22.9	100
	AS	12.5	2.1	**52.1**	33.3	100
	EU	12.5	3.8	7.5	**76.3**	100

AF, Africa; AM, America; AS, Asia; EU, Europe. Correct classifications are shown in bold.

Table 5. Discriminant functions with structural coefficients (SC) for each variable.

Statistics	Function 1	Function 2	Function 3
Eigenvalue	0.68	0.25	0.15
Wilks' Lambda	0.42	0.70	0.87
Chi-square	226.1	93.0	35.9
Proportion of trace (%)	63.0	23.1	13.9
Canonical correlation	0.63	0.45	0.36
p	<0.01	<0.01	<0.01
Team ball possessions	**−0.54**	**0.37**	0.16
Successful 2-point field goals	0.27	0.18	−0.09
Unsuccessful 2-point field goals	0.00	−0.04	−0.06
Successful 3-point field goals	0.22	0.06	0.12
Unsuccessful 3-point field goals	0.02	0.00	0.18
Successful free throws	−0.24	−0.16	**−0.41**
Unsuccessful free throws	**−0.47**	−0.09	**−0.76**
Defensive rebounds	−0.01	−0.20	0.25
Offensive rebounds	−0.09	0.02	−0.24
Assists	**0.38**	**0.39**	−0.16
Steals	−0.18	−0.18	0.01
Turnovers	**−0.36**	−0.09	0.05
Blocks	0.14	−0.04	0.07
Fouls committed	0.25	**−0.61**	**−0.40**

$|SC| \geq 0.30$ was considered relevant for discrimination (shown in bold).

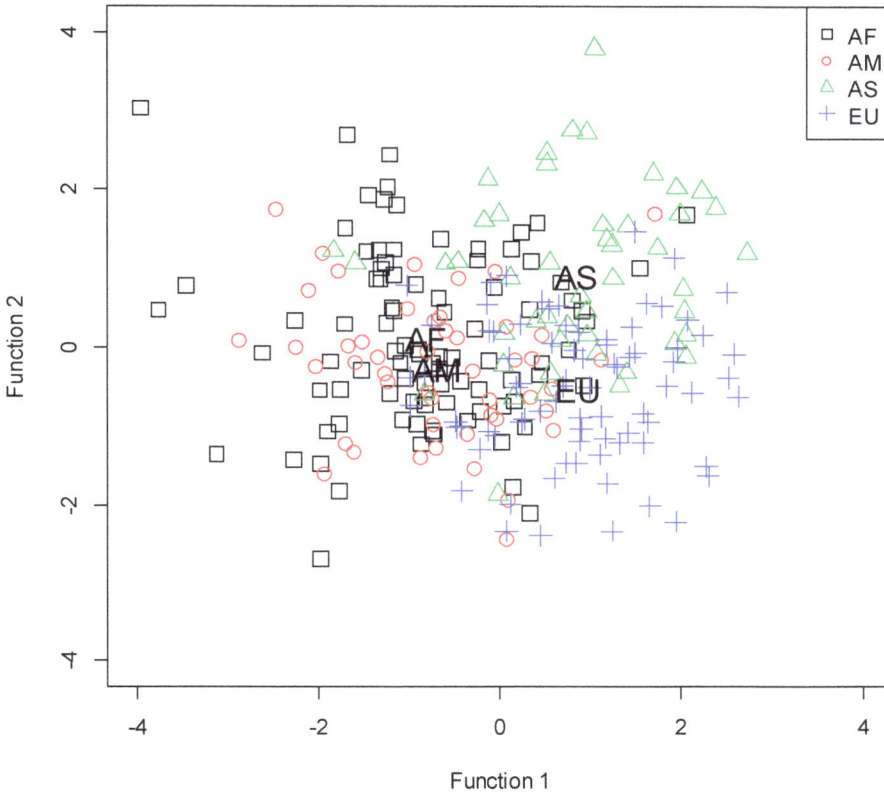

Figure 1. Territorial map of discriminant functions 1 and 2. AF, Africa; AM, America; AS, Asia; EU, Europe. Abbreviations plotted inside the figure indicate group centroids.

4. Discussion

This study analyzed game-related statistics of four continental championships for women held in 2017. The results showed that (a) significant differences among the continents were observed by ANOVA for 10 of 16 variables; (b) large effect size differences were observed for point difference, team ball possessions, unsuccessful free throws, assists, and fouls committed; (c) three significant functions that discriminate among the continents were obtained from the discriminant analysis. These results indicate that each continental championship has a specific performance profile and suggests that women's basketball games are played in a different manner in each region of the world.

The discriminant analysis showed that the correct classification rate for the European championship was the highest among the four continental championships. This result indicates a high homogeneity of the European championship. The mean point difference between the winning and losing teams in the European championship was the smallest among the four continental championships. In the latest FIBA World Ranking [8], a total of nine European countries, which is the highest number among the four continents, are listed in the top 20. It should be evident that the European championship was the most closely contested championship. One of the performance profiles of the European championship can be seen in ball possessions, which showed the lowest number among the four continental championships. A low number of possessions indicates that the game pace was slow, suggesting that European teams tended to run a set offense. This assumption is also supported by the fact that the number of assists, which has been considered as an indicator of a well-organized offense [16,27,28], in the European championship, was relatively high among the four continental championships (second to the Asian championship). These performance profiles were consistent with previously reported findings in continental championships for senior [2] and junior [3] men. It is suggested that the basic performance profiles of European basketball are common to both sexes.

In contrast to the European championship, the mean point difference between the winning and losing teams and the number of ball possessions in the African championship were the largest among the four continental championships. Among the 12 teams that took part in the African championship, the highest ranked team in the latest FIBA World Ranking was the 17th-ranked Senegal, whereas the lowest was the 75th-ranked Central African Republic [8]. This huge disparity among the participating teams was likely a cause of the large point difference between winning and losing teams in the African championship. Although a high number of possessions should result in a high number of offensive opportunities, the numbers of successful 2- and 3-point field goals in the African championship were relatively low among the four continental championships (the second lowest and the lowest, respectively). However, the numbers of free throws and turnovers in the African championship were relatively high among the four continental championships (the highest and the second highest, respectively). High numbers of free throws [2,3] and turnovers [2] in African games have also been reported in previous studies on men's championships. It is likely that players tended to lose possession before attempting a field goal or to be fouled during a shot in the African championship.

The correct classification rate for the Asian championship was relatively low among the four continental championships (52.1%, the second lowest), and 33.3% of the cases were misclassified into the European championship. This result indicates that a considerable number of the cases in the Asian championship have similar characteristics to the European championship. High numbers of successful field goals and assists, and low numbers of possessions, free throws and turnovers were common to the Asian and European championships. These findings were interesting because, unlike the European and African championships, the performance profiles of the Asian championship for women were largely different from Asian championships for senior [2] and junior [3] men. It has been suggested, because of a high number of possessions and a low number of assists, that the game pace is fast, and many points are scored after individual actions in Asian championships for men [2,3]. However, the present study showed that the Asian championship for women was characterized by a slow pace and well-organized offense. The difference in performance profiles between men and women may be related to differences in competitive performances in international competitions. As noted in the Introduction, four Asian countries are listed in the top 20 of the latest FIBA World Ranking for women [8], whereas only one country is listed in the top 20 of the FIBA World Ranking for men [7]. In addition, Asian women have shown better performances than Asian men in the Olympic Basketball Tournaments [29,30] and the FIBA World Cups [31,32].

The correct classification rate for the American championship was the lowest (43.8%) among the four continental championships. In the American championship, 29.2% of the cases were misclassified into the African championship, and 22.9% of the cases were misclassified into the European championship. These results indicate that the homogeneity of the American championship was low, and some of the cases have similarities with the African championship, whereas other cases have similarities with the European championship. Although it is difficult to clarify the performance profiles of the American championship due to the low homogeneity, some characteristics specific to the American championship could be found in game-related statistics. The numbers of points scored, successful 2-point field goals, and assists in the American championship were the lowest among the four continental championships, whereas the numbers of steals and turnovers in the American championship were the highest among the four continental championships. These findings were inconsistent with previously reported findings in continental championships for men held in 2015 [2], in which the numbers of points scored and successful 2-point field goals in the American championship were the highest among the four continental championships. In contrast to the American championship for men, the American championship for women seems to be a defense-oriented, low-scoring championship.

Although this study provides novel information that each continental championship has a specific performance profile, it is not without limitations. Since the methodology is purely quantitative in nature, qualitative elements of the game, such as types of offense [33,34] and defense [35,36], remain unrevealed. Future studies on qualitative elements of the game would compensate for this limitation and provide further understanding of regional differences in women's basketball.

From a practical perspective, this study will help players and coaches of women's national teams prepare for international competitions. At international competitions, national teams are required to play games against relatively unfamiliar teams in a short period of time. Detailed information about opponent teams can only be obtained through specific scouting of each opponent. However, basic information about opponent teams can be obtained from this study based on the region of the world where each opponent belongs.

5. Conclusions

This study identified regional differences in basketball games among four continental championships for women held in 2017. The Asian and European championships overall showed similar performance profiles: Low numbers of possessions and turnovers, and high numbers of successful field goals and assists. However, the European championship was more closely contested than the Asian championship. The African championship was characterized by high numbers of possessions, free throws, and turnovers, and a low number of successful field goals. The homogeneity of the American championship was low, and some of the cases have similarities with the African championship, whereas other cases have similarities with the European championship. On average, the American championship was characterized by low numbers of successful field goals and assists, and high numbers of steals and turnovers. It is suggested that women's basketball games are played in a different manner in each region of the world.

Funding: This research was funded by JSPS KAKENHI, grant number 18K10837.

Conflicts of Interest: The author declares no conflict of interest.

References

1. International Basketball Federation (FIBA). Presentation, Facts & Figures. Available online: http://www.fiba.basketball/presentation#tab=element_2_1 (accessed on 22 June 2018).
2. Ibáñez, S.J.; González-Espinosa, S.; Feu, S.; García-Rubio, J. Basketball without borders? Similarities and differences among Continental Basketball Championships. *RICYDE. Rev. Int. Cienc. Deporte* **2018**, *14*, 42–54. [CrossRef]
3. Madarame, H. Are regional differences in basketball already established in under-18 games? *Motriz Rev. Educ. Fis.* **2018**, in press.
4. Sampaio, J.; Godoy, S.I.; Feu, S. Discriminative power of basketball game-related statistics by level of competition and sex. *Percept. Mot. Skills* **2004**, *99*, 1231–1238. [CrossRef] [PubMed]
5. Gómez, M.A.; Lorenzo, A.; Ibáñez, S.J.; Sampaio, J. Ball possession effectiveness in men's and women's elite basketball according to situational variables in different game periods. *J. Sports Sci.* **2013**, *31*, 1578–1587. [CrossRef] [PubMed]
6. Madarame, H. Age and sex differences in game-related statistics which discriminate winners from losers in elite basketball games. *Motriz Rev. Educ. Fis.* **2018**, *24*, e1018153. [CrossRef]
7. International Basketball Federation (FIBA). FIBA World Ranking presented by Nike, Men. Available online: http://www.fiba.com/rankingmen (accessed on 20 June 2018).
8. International Basketball Federation (FIBA). FIBA World Ranking presented by Nike, Women. Available online: http://www.fiba.basketball/rankingwomen (accessed on 20 June 2018).

9. Conte, D.; Lukonaitiene, I. Scoring strategies differentiating between winning and losing teams during FIBA EuroBasket Women 2017. *Sports (Basel)* **2018**, *6*, 50. [CrossRef] [PubMed]
10. Leicht, A.; Gomez, M.; Woods, C. Team performance indicators explain outcome during women's basketball matches at the Olympic Games. *Sports (Basel)* **2017**, *5*, 96. [CrossRef] [PubMed]
11. Madarame, H. Defensive rebounds discriminate winners from losers in European but not in Asian women's basketball championships. *Asian J. Sports Med.* **2018**, *9*, e67428. [CrossRef]
12. Gómez, M.A.; Lorenzo, A.; Sampaio, J.; Ibáñez, S.J. Differences in game-related statistics between winning and losing teams in women's basketball. *J. Hum. Mov. Stud.* **2006**, *51*, 357–369.
13. Moreno, E.; Gómez, M.A.; Lago, C.; Sampaio, J. Effects of starting quarter score, game location, and quality of opposition in quarter score in elite women's basketball. *Kinesiology* **2013**, *45*, 48–54.
14. Şentuna, M.; Şentuna, N.; Özdemir, N.; Serter, K.; Özen, G. The investigation of the effects of some variables in the playoff games played in Turkey Women's Basketball Super League between 2013–2017 on winning and losing. *Phys. Educ. Stud.* **2018**, *22*, 146–150. [CrossRef]
15. Gómez, M.A.; Lorenzo, A.; Ortega, E.; Sampaio, J.; Ibáñez, S.J. Game related statistics discriminating between starters and nonstarters players in Women's National Basketball Association League (WNBA). *J. Sports Sci. Med.* **2009**, *8*, 278–283.
16. Garcia, J.; Ibáñez, S.J.; De Santos, R.M.; Leite, N.; Sampaio, J. Identifying basketball performance indicators in regular season and playoff games. *J. Hum. Kinet.* **2013**, *36*, 161–168. [CrossRef] [PubMed]
17. Paulauskas, P.; Masiulis, N.; Vaquera, A.; Figueira, B.; Sampaio, J. Basketball game-related statistics that discriminate between European players competing in the NBA and in the Euroleague. *J. Hum. Kinet.* **2018**, in press.
18. International Basketball Federation. *FIBA Statisticians' Manual 2016*; FIBA: Mie, Switzerland, 2016.
19. Sampaio, J.; Lago, C.; Drinkwater, E.J. Explanations for the United States of America's dominance in basketball at the Beijing Olympic Games (2008). *J. Sports Sci.* **2010**, *28*, 147–152. [CrossRef]
20. Ibáñez, S.J.; García-Rubio, J.; Gómez, M.A.; Gonzalez-Espinosa, S. The impact of rule modifications on elite basketball teams' performance. *J. Hum. Kinet.* **2018**, in press.
21. Gómez, M.A.; Avugos, S.; Ángel Oñoro, M.; Lorenzo Calvo, A.; Bar-Eli, M. Shaq is not alone: Free-throws in the final moments of a basketball game. *J. Hum. Kinet.* **2018**, *62*, 135–144. [CrossRef] [PubMed]
22. Sampaio, J.; Janeira, M. Statistical analyses of basketball team performance: understanding teams' wins and losses according to a different index of ball possessions. *Int. J. Perform. Anal. Sport* **2003**, *3*, 40–49. [CrossRef]
23. Oliver, D. Watching a game: Offensive score sheets. In *Basketball on Paper: Rules and Tools for Performance Analysis*; Potomac Books: Washington, DC, USA, 2004; pp. 8–28.
24. R Core Team. *R: A Language and Environment for Statistical Computing*; R Foundation for Statistical Computing: Vienna, Austria, 2018.
25. Cohen, J. A power primer. *Psychol. Bull.* **1992**, *112*, 155–159. [CrossRef] [PubMed]
26. Madarame, H. Game-related statistics which discriminate between winning and losing teams in Asian and European men's basketball championships. *Asian J. Sports Med.* **2017**, *8*, e42727. [CrossRef]
27. Lorenzo, A.; Gómez, M.A.; Ortega, E.; Ibáñez, S.J.; Sampaio, J. Game related statistics which discriminate between winning and losing under-16 male basketball games. *J. Sports Sci. Med.* **2010**, *9*, 664–668. [PubMed]
28. Gómez, M.A.; Ibáñez, S.J.; Parejo, I.; Furley, P. The use of classification and regression tree when classifying winning and losing basketball teams. *Kinesiology* **2017**, *49*, 47–56. [CrossRef]
29. International Olympic Committee (IOC). Basketball Women. Available online: https://www.olympic.org/basketball/basketball-women (accessed on 25 June 2018).
30. International Olympic Committee (IOC). Basketball Men. Available online: https://www.olympic.org/basketball/basketball-men (accessed on 25 June 2018).
31. International Basketball Federation (FIBA). All Time Medalists: Results of the 17 editions of "FIBA World Championship for Women". Available online: http://www.fiba.com/world/women/2014/alltimemedalists (accessed on 25 June 2018).
32. International Basketball Federation (FIBA). All Time Medalists: Results of the 18 editions of "FIBA Basketball World Cup". Available online: http://www.fiba.com/basketballworldcup/2019/alltimemedalists (accessed on 25 June 2018).
33. Ciampolini, V.; Ibáñez, S.J.; Nunes, E.L.G.; Borgatto, A.F.; Nascimento, J.V.D. Factors associated with basketball field goals made in the 2014 NBA finals. *Motriz Rev. Educ. Fis.* **2017**, *23*, e1017105. [CrossRef]

34. Conte, D.; Favero, T.G.; Niederhausen, M.; Capranica, L.; Tessitore, A. Determinants of the effectiveness of fast break actions in elite and sub-elite Italian men's basketball games. *Biol. Sport* **2017**, *34*, 177–183. [CrossRef] [PubMed]

35. Gómez, M.A.; Lorenzo, A.; Ibáñez, S.J.; Ortega, E.; Leite, N.; Sampaio, J. An analysis of defensive strategies used by home and away basketball teams. *Percept. Mot. Skills* **2010**, *110*, 159–166. [CrossRef] [PubMed]

36. Gómez, M.A.; Tsamourtzis, E.; Lorenzo, A. Defensive systems in basketball ball possessions. *Int. J. Perform. Anal. Sport* **2006**, *6*, 98–107.

© 2018 by the author. Licensee MDPI, Basel, Switzerland. This article is an open access article distributed under the terms and conditions of the Creative Commons Attribution (CC BY) license (http://creativecommons.org/licenses/by/4.0/).

![sports logo] *sports*

MDPI

Article

Variability of Jump Kinetics Related to Training Load in Elite Female Basketball

Jan Legg [1,2,*], David B. Pyne [1,2] ![ORCID], Stuart Semple [2] and Nick Ball [2]

[1]   Strength and Conditioning, Australian Institute of Sport, Bruce 2617, Australia; david.pyne@ausport.gov.au
[2]   Research Institute for Sport and Exercise, University of Canberra, Bruce 2617, Australia;
       Stuart.Semple@canberra.edu.au (S.S.); Nick.Ball@canberra.edu.au (N.B.)
*   Correspondence: jan.legg@ausport.gov.au; Tel.: +61-2-423-232-047

Received: 31 August 2017; Accepted: 2 November 2017; Published: 4 November 2017

Abstract: The purpose of this study was to quantify changes in jump performance and variability in elite female basketballers. Junior and senior female representative basketball players ($n = 10$) aged 18 ± 2 years participated in this study. Countermovement jump (CMJ) data was collected with a Gymaware™ optical encoder at pre-, mid-, and post-season time points across 10 weeks. Jump performance was maintained across the course of the full season (from pre to post). Concentric peak velocity, jump height, and dip showed the most stability from pre- to post-season, with the %CV ranging from 5.6–8.9%. In the period of the highest training load (mid-season), the variability of within-subject performance was reduced by approximately 2–4% in all measures except for jump height. Altered jump mechanics through a small (0.26 effect size) increase in dip were evident at mid-season, suggesting that CMJ analysis is useful for coaches to use as an in-season monitoring tool. The highest coefficient of variation (8–22%CV) in inter-set scores in all measures except eccentric peak velocity also occurred mid-season. It appears that in-season load not only impairs jump performance, but also movement variability in basketball players.

Keywords: countermovement jump; variability; basketball

1. Introduction

Vertical jump performance has been studied extensively in male basketball players as an indicator of lower limb power, with more elite players recording greater jump heights [1]. Time on court correlates highly with the anaerobic performance of vertical jump height, speed, and agility, indicating that physical capabilities play a strong role in team selection [2]. The countermovement jump (CMJ) is considered a practical assessment tool in elite sports to examine kinetic and or kinematic performance variables [3,4].

The ability to produce force is essential for jumping ability in basketball players, with jumping considered an acceptable measure for evaluating the stretch shortening cycle (SSC) [5,6]. Muscle function and the ability to quickly transition from eccentric to concentric contractions via the SSC is critical to many offensive and defensive manoeuvres performed in basketball, including rebounding, shooting, and sprinting. Analysis of an athlete's SSC can also be useful in monitoring the effects of fatigue on performance [7]. This type of analysis has not been conducted previously on elite female basketball players across a competition season.

The use of a linear position transducer (LPT) by high-performance coaches and their support staff to measure CMJ performance is increasingly common. An LPT provides a portable and effective method of analysing the displacement of the bar or body [8]. Kinematic data from LPTs is differentiated to estimate force and power when subject mass is factored in, with strong relative validity compared with a force plate [8,9]. Until recently, analysis of CMJ data has been limited to values relating to the concentric phase of the jump, such as jump height and peak power. However, the importance of monitoring eccentric jump variables as a result of altered jumping mechanics after training adaptations or fatigue is now recognized [10,11]. CMJ testing can also provide valuable insights into neuromuscular fatigue, response to training loads, and subsequent recovery in high-performance sport environments [3]. This information can be collated with subjective internal training load data to give a comprehensive picture of the athlete's preparation and response to training [12].

Countermovement jump analysis can provide worthwhile information, but it is imperative that coaches understand the typical variation or repeatability of the testing. It is critical for coaches to understand where a meaningful change in performance has occurred or whether the magnitude of changes lie within the normal reproducibility of the outcome variable (jump performance), known as the typical error [13]. The level of expertise an athlete has acquired in the performance skill will affect reliability, with elite athletes having a demonstrated ability to achieve outcome performances more consistently [14]. This is despite dynamical systems theory and coordination profiling that proposes that as an athlete's skill level and expertise improves the motor system variability will increase in order to achieve these consistent performance outcomes [15]. These theories propose that skilled performers adapt to their surroundings and any unique constraints (environmental, task, and organismic) to achieve stable task execution through movement variability. This adaptation mechanism may only be applicable to open sport skills such as shooting in basketball when under pressure from an opposing player rather than the closed skill of a CMJ, particularly in developing athletes [16,17]. Consequently, whilst kinetic variability is likely to be relatively stable in a CMJ, this has yet to be demonstrated.

Acceptable within-subject reliability for force-related measures of the CMJ has been defined as a coefficient of variation (%CV) between 2.8 and 9.5% for single trials, and between 0.8 and 6.2% when six trials are used [3,18]. The error rate for a six-jump protocol is estimated between 1.1–3.2% for most kinetic and kinematic variables, except for rate of force development, which may be as high as 13–16% [3]. While studies have focused on the magnitude of training interventions [10,19], the changes in within-subject reliability of the kinetic profile from the CMJ in elite female basketball players across the course of a training period is unclear. Research is required to determine if kinetic variability can be decreased as a function of improved performance in basketball players.

The purpose of this study was to quantify the pattern of within-subject variability in jump height and kinetic variables in elite female basketball players over the course of a competitive season. This investigation also sought to determine if performance increases in jump height and kinetic variables were evident within the 10-week in-season training period. This information should provide valuable insight to coaches by highlighting the specific areas athletes can target in training to improve power production and jump performance. We hypothesized that while jump performance may not improve substantially in-season, the within-subject variability of SSC parameters would likely decrease as a result of structured training throughout the competition phase.

2. Materials and Methods

2.1. Experimental Approach to the Problem

Ten female Basketball Australia Centre of Excellence scholarship holders completed a ten-week in-season strength and conditioning program to improve jump technique and performance. A single-group longitudinal assessment was employed over the competition season. Kinetic variables were recorded from the CMJ at three different time points across the season (pre-, mid-, and post-season) with within-subject and between-subject reliability analysis undertaken to assess kinetic variability.

2.2. Participants

Australian female junior and senior representative basketball players ($n = 10$) aged 18 ± 2 years, height 1.85 ± 0.09 m, mass 75.2 ± 7.2 kg, sum of seven skinfolds 82.6 ± 13.9 mm participated in this study. All players performed jump testing as part of their usual training program during a designated afternoon training session. All participants trained together and represented the same team throughout the competition. Informed consent was given by all participants, with ethics approval granted through the Australian Institute of Sport Ethics Committee. Players were excluded from individual testing sessions if they were not currently completing a full training load for basketball due to injury, fatigue, or illness. The participants had a range of resistance training experience with 12 months ($n = 6$), 2 years ($n = 2$), and 6 years ($n = 2$).

2.3. Procedures

Resistance Training and Training Load. Athletes were required to complete their usual periodized resistance training sessions consisting of 2–3 full body sessions per week (depending on scheduling of games). The distribution of exercise selection and consequent loading across each training week was stability and control 20%, power 30–40%, and strength 40–50%. Mean weekly training load consisted of three full-team on-court training sessions of 2–2.5 h duration, two individual high-intensity on-court training sessions of 30 min duration, and two low-intensity 60 min shooting sessions. This load was elevated at the mid-season point during an intensive training camp of 5 days duration to a mean of 4 h of court-based team training daily—approximately a two-fold increase (doubling) in duration. Rating of perceived exertion (RPE) was collected following each training session as well as games on the Borg 10 point scale [20]. Athletes were required to identify the RPE for each session, and this was multiplied by the session duration. The total training load using the session-RPE method for each week at the pre-, mid-, and post-season was then calculated as the total load sum of sessional data for each athlete aggregated as a group mean. Match loads were included in the total training load as minutes played multiplied by the RPE score. All participants were familiar with using RPE as part of their regular load monitoring of training and games.

Jump Tests. Countermovement jump data was collected with a Gymaware™ optical encoder (50 Hz sample period with variable rate sampling with level crossing detection to capture data points; Kinetic Performance Technology, Canberra, Australia) [21]. The Gymaware™ optical encoder was attached via a tether to the right side of a 0.3 kg wooden bar. Bar placement on the back for each subject was between the superior portion of the scapula and vertebra C7. The stance for each subject was constrained to within 15 cm of the lateral portion of the individual's deltoid, as specified by McBride et al. [22]. Participants initiated the CMJ via a downward countermovement to a self-selected depth, followed immediately by a maximal effort vertical jump. Participants were instructed to keep constant downward pressure on the bar throughout the jump and encouraged to reach a maximum jump height with every trial in an attempt to maximize power output. Participants were encouraged and reminded between trials to "jump high" and "jump fast". Each subject performed five trials, with a small pause between each trial to steady themselves and stand tall. The following variables were collected for normalised power: peak power (W) and concentric (Conc) mean power (W). Mean watts (W/kg) and peak watts (W/kg) were used to determine absolute power. Concentric peak velocity (m/s), eccentric peak velocity (m/s), as well as jump height (cm) and dip (cm) were also recorded.

Prior to jump testing, all athletes completed the same 10-min warm up consisting of dynamic flexibility work followed by 10 bodyweight squats, three bodyweight CMJs at subjective intensity of 60% maximal effort, and three bodyweight CMJs at 90% maximal effort. All athletes had previously completed this type of jump testing as part of their regular training for at least 4 weeks prior to the investigation.

2.4. Statistical Analyses

Means and standard deviations (SD) were calculated for the kinetic and kinematic variables at pre-season, mid-season, and post season. Independent variables were the time point of the testing (pre/mid/post), while the dependent variables were power, velocity, jump height, and dip. The effect of training on jump performance was quantified by determining the mean change in test scores over the 10-week training period using an independent student's *t*-test. Precision of estimation was expressed as the 90% confidence limits. A standardized mean effect was used to characterize the magnitude of change. Magnitudes of standardized effects were interpreted against the following criteria: trivial 0.0, small 0.2, moderate 0.6, and large 1.0 [13]. An effect was deemed unclear if its confidence limits simultaneously overlapped the thresholds for a substantially positive and negative change.

To examine the mean within-subject variability in jump performance and kinetics at baseline and changes in variability across the season, we computed the % coefficient of variation (%CV). To compare the magnitude of change in variation for phase of the season we divided the %CV for consecutive pairs of testing sessions. Ratios within a range of 0.87 to 1.15 were considered trivial; a ratio >1.15 indicated that CMJ performance was substantially more variable, whereas a ratio <0.87 indicated that test results were substantially less variable [23].

To examine the mean between-subject variability in jump performance and kinetics at baseline and changes in variability across the season, we calculated an intraclass correlation coefficient (ICC). We interpreted the magnitude of the correlation using the thresholds of 0.20 (low), 0.50 (moderate), 0.75 (high), 0.90 (very high), and 0.99 (extremely high) consistency [13]. A difference >0.10 in the correlation coefficient between time points across the season was deemed substantial.

3. Results

No substantial changes in jump performance occurred from pre- to post-season (Table 1), with jump height performance measures stable at all measurement points across the season. A moderate reduction in watts/kg and a small reduction in concentric mean power was evident from the pre-season to mid-season. However, these effects were reversed by the end of the season, with a moderate and small increase in these measures from the mid-season to post-season. There were decreased values for %CV, suggesting more consistent jumping, from the beginning of the season to mid-season (which coincides with the period of the highest training loads) in all variables except jump height.

Table 2 details the inter-set reliability across the season, with mid-season showing the highest variation in executing five CMJ in all variables except eccentric peak velocity, dip, and height, which remained stable. Eccentric peak velocity also had a large increase in variability in the post-season testing, exhibiting a within-set %CV of 46%; however, this change was deemed unclear. There were substantial improvements in within-set reliability for mean watts (W/kg) and mean power from pre-season to post-season testing. Concentric peak velocity, jump height, and dip showed the greatest consistency from pre- to post-season, with the CV ranging from 6–10%. However concentric peak velocity was less reliable mid-season, with an increase in %CV and a substantial reduction in reliability from moderate to low.

Group training loads were recorded to be the following from RPE scores: pre-season 3195 ± 1083 (arbitrary units; mean ± SD), mid-season 4344 ± 1376, and post-season 2161 ± 1043. The groups' training loads were markedly higher during mid-season, corresponding with high reported RPE values.

Table 1. Changes in within-subject lower body power characteristics in women's basketball players across a season using a countermovement jump. %CV coefficient of variation; pre-season; mid-season, post-season. ΔX: change in the mean; mod: moderate; Ecc: eccentric; Conc: concentric.

	Pre-Season		Mid-Season		Change (Pre vs. Mid)		Post-Season		Change (Pre vs. Post)		Change (Mid vs. Post)	
	Mean	%CV	Mean	%CV	ΔX	Effect Size	Mean	%CV	ΔX	Effect Size	ΔX	Effect Size
Mass (kg)	76.2 ± 7.6		76.3 ± 8.3				76.4 ± 7.8					
Mean Watts (W/kg)	43.0 ± 5.1	12	41.1 ± 4.4	11	−2.0	−0.36 mod	43.2 ± 7.2	17	0.17	0.00 trivial	2.13	0.35 mod
Peak Power (W)	5130 ± 862	17	5160 ± 651	13	30	0.06 trivial	5039 ± 889	18	−91	−0.10 trivial	−121	−0.16 trivial
Peak Watts (W/kg)	67.2 ± 9.6	14	67.8 ± 6.8	10	0.18	0.05 trivial	66.3 ± 12.2	18	−2.0	−0.17 trivial	−1.5	−0.22 small
Conc Mean Power (W)	3275 ± 592	18	3129 ± 422	14	−171	−0.26 small	3269 ± 499	15	−30	−0.03 trivial	140	0.23 small
Conc Peak Velocity (m/s)	3.22 ± 0.25	8	3.23 ± 0.19	6	−0.02	−0.05 trivial	3.23 ± 0.24	7	−0.02	−0.08 trivial	0.01	−0.03 trivial
Height (m)	0.39 ± 0.05	13	0.40 ± 0.07	17	0.004	0.00 trivial	0.39 ± 0.06	15	−0.01	−0.18 trivial	−0.01	−0.18 trivial
Dip (m)	0.43 ± 0.11	27	0.46 ± 0.11	25	0.013	0.26 small	0.44 ± 0.12	29	−0.004	−0.02 trivial	−0.02	−0.21 small
Ecc Peak Velocity (m/s)	1.0 ± 0.2	23	0.9 ± 0.2	19	−0.07	−0.09 trivial	1.0 ± 0.2	22	−0.05	−0.04 trivial	0.02	0.05 trivial

Table 2. Inter-set reliability of performing five countermovement jumps (CMJs) at each time point presented as %CV, 90% confidence interval, and the intraclass correlation coefficient (ICC).

	Pre-Season	ICC	Mid-Season	ICC	Post-Season	ICC
Mean Watts (w/kg)	12, 9–22	0.46	14, 11–27	0.29	9, 7–17	0.69
Peak Power (W)	15, 11–27	0.62	18, 14–34	0.22	15, 12–28	0.50
Peak Watts (w/kg)	15, 11–27	0.45	18, 14–34	0.05	15, 12–28	0.54
Mean Power (W)	12, 9–22	0.64	14, 11–27	0.46	9, 7–17	0.69
Conc Peak Velocity (m/s)	6, 5–11	0.62	8, 6–15	0.17	6, 4–10	0.60
Height (cm)	7, 5–11	0.79	8, 6–15	0.84	6, 5–11	0.87
Dip (cm)	9, 7–15	0.93	10, 7–17	0.89	7, 5–12	0.94
Eccentric Peak Velocity (m/s)	27, 24–67	0.58	22, 18–50	0.59	46, 43–138	0.43

Thresholds for ICC: 0.20 (low), 0.50 (moderate), 0.75 (high), 0.90 (very high), and 0.99 (extremely high) consistency.

4. Discussion

Training during a ten-week in-season competition phase did not elicit substantial changes in jump performance in a cohort of elite female basketball players. While kinetic performance of relative (moderate) and mean (small) concentric power declined from pre- to mid-season, these measures were restored by the post-season. Consistency of jumping (reliability) in the context of performing a standard five jump CMJ protocol was improved across the course of the season for mean watts/kg, mean power, concentric peak velocity, jump height, and dip. These outcomes indicate that while performance markers such as height and power did not improve, consistency of jump performance improved as a result of a structured training program throughout the competition phase. Within-subject variation decreased across the season in mean power and dip, whereas height and concentric peak velocity remained steady, demonstrating increased reliability as well as consistency in the primary performance outcomes.

The ability to efficiently produce force is essential for jumping ability for basketball players, with the propulsive action considered to be an acceptable measure for evaluating explosive characteristics [5,6]. In the protocols used in this investigation (i.e., restricting arm swing), jump height values are lower than those reported in other investigations that permitted a more natural jumping action [1]. Concentric peak velocity, jump height, and dip showed the greatest consistency from pre- to post-season, with the CV ranging from 6–10%, which may be attributed to the established relationship between jump height and velocity [24]. The concentric peak velocity of the players in the pre-season of this investigation (3.22 ± 0.25 m/s) was comparable to other sports of similar age and performance (women's football 3.00 ± 0.20m/s and netball 2.80 ± 0.20 m/s) [25]. Concentric peak velocity was less reliable mid-season, with an increase in the CV% and a marked reduction in the intra-set consistency from moderate to low reliability. This reduction may relate to altered jumping mechanics of the athletes during this fatigued training period, as dip was increased to maintain jump height despite more variable velocity measures. The altered jumping mechanics may also explain the large changes in the power characteristics observed in the peak power, peak watts, and concentric mean power as the athletes modified their jump in order to achieve the performance outcome of jump height.

The improved consistency of jumping in this squad may be attributed to improvements in coordination, control and skill as the season progressed. The likely explanation is improvement in jumping motor patterns. Given the limited experience of strength training in the majority of the squad, motor learning as described by Newell would most likely have occurred throughout the competition phase [26]. Strength and power levels can increase significantly for athletes trained in a supervised environment, when compared to those training in an unsupervised environment [27]. An increase in motor control and coordination of jumping skills would be anticipated in athletes training for the first time within a centralized training environment under direct coaching and supervision [28]. It is likely that the athletes in this investigation were beginning to master the skill of jumping, as demonstrated by the ability to perform more consistently [14]. This outcome follows traditional motor learning principles that once skilled performance has been acquired there is likely a concomitant reduction in coordination variability. However, the changed dip patterns seen in this squad following fatigue raises the question of whether coordination variability in athletes is influenced by both adaptation and fatigue.

With increased training loads during mid-season (4344 $\pm$ 1376) compared to pre-season (3195 $\pm$ 1083), the magnitude of the dip was increased across the squad, suggesting that the athletes were completing a deeper squat prior to commencing the jump in order to generate the necessary power to achieve a similar jump height. This outcome may be explained by the concept that skilled performers are able to demonstrate increased movement variability and altered movement strategies to achieve consistent performance outcomes [5,11]. Impulse was likely to have been altered in this fatigued state, with time under the force-time curve increased [29]. This effect is likely related to athletes inherently trying to generate more force through increased eccentric loading when in a fatigued state, although only a trivial positive increase in peak power and peak watts was observed from pre- to mid-season. Elite snowboard cross-athletes exhibited a similar change in jump mechanics when athletes were fatigued with increased dip in an effort to maintain jump height [11]. While neuromuscular fatigue appears to have altered the biomechanics of the CMJ through increased dip measures in both Gathercoleet al.'s research and this investigation, no significant decreases in capacity as measured by jump height were evident [11]. This outcome may be explained by the motor learning concept that in well-learned movements, consistent performance outcome is often associated with high intra-limb joint coordination variability whereby the movement pattern an athlete uses may appear different but the end result is the same [30].

When training loads based on RPE were at their highest point (mid-season), the within-group variation in selected kinetic variables was reduced. It might be the case that neuromuscular fatigue limits the jumping capacity of the best overall jumpers and brings them to similar levels of lesser-skilled jumpers within the group. Neural control as demonstrated through coordination could be compromised following fatigue and affect the strength available for optimal jump performance [31]. The change in mean scores of peak power was increased mid-season in this squad, along with small (-0.36 and -0.26) decrements of the within-subject strength-based variables of relative peak watts (W/kg) and concentric mean power. Over the course of this investigation, strength markers were not able to be collected, but these results are likely related to a reduction in maximum strength driven by neuromuscular fatigue at the mid-point of the season. A reduction in strength has been linked to decreased jumping ability in numerous sports [22,32].

This investigation did not show any significant changes in jump performance within the 10-week in-season training block. The stability of jump height in this in this study is in contrast to a 7-week in-season competition phase for junior male rugby league players that detailed accumulating fatigue and reduced CMJ performance [7]. Reductions in CMJ performance was also reported in division 1 collegiate male soccer players who exhibited a 13.8% reduction in vertical jump across an 11-week season [33]. While fatigue was evident in this investigation at the mid-season point as demonstrated by impaired jump mechanics, recovery was sufficient throughout the season to avoid any significant decreases in jump performance.

5. Conclusions

During the course of a competitive season, basketball players are exposed to a combination of the rigors of games and training demands both on the court and in the weight room. Strength and conditioning coaches should consider monitoring movement variability of the CMJ to assess training effects as well as the degree of neuromuscular fatigue. This investigation showed increased movement variability in the CMJ when training loads were at their peak along with increased fatigue. While the commonly investigated performance marker of jump height appears stable even when players are in a fatigued state, strength and conditioning coaches can monitor eccentric jump variables for signs of overtraining or overreaching. An increase in eccentric duration has the potential to negatively affect sport outcomes. For example, a longer duration spent achieving maximal push off on the basketball court could lead to a missed pass or rebound. Given the importance of movement speed and mechanical efficiency, these results could affect the training load prescribed to athletes during a competition season.

Author Contributions: Jan Legg and Nick Ball conceived and designed the experiment. Jan Legg performed the experiment and Jan Legg and David Pyne analyzed the data. Jan Legg wrote the paper and David Pyne, Nick Ball and Stuart Semple assisted in editing the paper.

Conflicts of Interest: The authors declare no conflict of interest.

References

1. Ziv, G.; Lidor, R. Vertical jump in female and male basketball players—A review of observational and experimental studies. *J. Sci. Med. Sport* **2010**, *13*, 332–339. [CrossRef] [PubMed]
2. Hoffman, J.R.; Tenenbaum, G.; Maresh, C.M.; Kraemer, W.J. Relationship between athletic performance tests and playing time in elite college basketball players. *J. Strength Cond. Res.* **1996**, *10*, 67–71.
3. Taylor, K.-L.; Cronin, J.; Gill, N.D.; Chapman, D.W.; Sheppard, J. Sources of variability in iso-inertial jump assessments. *Int. J. Physiol. Perform.* **2010**, *5*, 546–558.
4. Cormack, S.J.; Newton, R.U.; Mcguigan, M.R.; Cormie, P. Neuromuscular and Endocrine Responses of Elite Players During an Australian Rules Football Season. *Int. J. Sports Physiol. Perform.* **2008**, *3*, 439–453. [CrossRef] [PubMed]
5. Markovic, G.; Dizdar, D.; Jukic, I.; Cardinale, M. Reliabiliy and factorial validity of squat and coutermovement jump tests. *J. Strength Cond. Res.* **2004**, *18*, 551–555. [PubMed]
6. Young, W.; Cormack, S.; Crichton, M. Which jump variables should be used to assess explosive leg muscle function? *Int. J. Sports Physiol. Perform.* **2011**, *6*, 51–57. [CrossRef] [PubMed]
7. Oliver, J.L.; Lloyd, R.S.; Whitney, A. Monitoring of in-season neuromuscular and perceptual fatigue in youth rugby players. *Eur. J. Sports Sci.* **2015**, *15*, 514–522. [CrossRef] [PubMed]
8. Crewther, B.T.; Kilduff, L.P.; Cunningham, D.J.; Cook, C.; Owen, N.; Yang, G.Z. Validating Two Systems for Estimating Force and Power. *Int. J. Sports Med.* **2011**, *32*, 254–258. [CrossRef] [PubMed]
9. Cronin, J.B.; Hing, R.D.; McNair, P.J. Reliability and validity of a linear position transducer for measuring jump performance. *J. Strength Cond. Res.* **2004**, *18*, 590–593. [PubMed]
10. Cormie, P.; McBride, J.M.; McGaulley, G.O. Power-time, foce-time, and velocity-time curve analysis of the countermovement jump: impact of training. *J. Strength Cond. Res.* **2009**, *23*, 177–186. [CrossRef] [PubMed]
11. Gathercole, R.; Sporer, B.; Stellingwerff, T.; Sleivert, G. Alternative Countermovement-Jump Analysis to Quantify Acute Neuromuscular Fatigue. *Int. J. Sports Physiol. Perform.* **2015**, *10*, 84–92. [CrossRef] [PubMed]
12. Scanlan, A.T.; Wen, N.; Tucker, P.S.; Dalbo, V.J. The Relationships Between Internal and External Training Load Models During Basketball Training. *J. Strength Cond. Res.* **2014**, *28*, 2397–2405. [CrossRef] [PubMed]
13. Hopkins, W.G. Measures of reliability in sports medicine and science. *Sports Med.* **2000**, *30*, 1–15. [CrossRef] [PubMed]
14. Seifert, L.; Button, C.; Davids, K. Key Properties of Expert Movement Systems in Sport. *Sports Med.* **2013**, *43*, 167–178. [CrossRef] [PubMed]
15. Davids, K.; Araujo, D.; Vilar, L.; Renshaw, I.; Pinder, R.A. An ecological dynamics approach to skill acquisition: Implications for development of talent in sport. *Talent Dev. Excell.* **2013**, *5*, 21–34.
16. Colley, A.M.; Beech, J.R. *Cognition and Action in Skilled Behaviour*; Elsevier: Amsterdam, The Netherlands, 1988; Volume 55.
17. Jansen, J.J.; van den Bosch, F.A.; Volberda, H.W. Exploratory innovation, exploitative innovation, and performance: Effects of organizational antecedents and environmental moderators. *Manag. Sci.* **2006**, *52*, 1661–1674. [CrossRef]
18. Cormack, S.J.; Newton, R.U.; Mcguigan, M.R.; Doyle, T.L. Reliability of Measures Obtained During Single and Repeated Countermovement Jumps. *Int. J. Sports Physiol. Perform.* **2008**, *3*, 131–144. [CrossRef] [PubMed]
19. Argus, C.K.; Gill, N.D.; Keogh, J.W.; Mcguigan, M.R.; Hopkins, W.G. Effects of two Contrast Training Programs on Jump Performance in Rugby Union Players During a Competition Phase. *Int. J. Sports Physiol. Perform.* **2012**, *7*, 68–75. [CrossRef] [PubMed]

20. Borg, G. *Borg's Perceived Exertion and Pain Scales*; Human kinetics: Champaign, IL, USA, 1998.

21. KineticPerformance. Kinetic Performance: GymAware Sampling Method. 2017. Available online: https://kinetic.com.au/pdf/sample.pdf (accessed on 1 November 2017).

22. McBride, J.M.; Jeffrey, M. A comparison of strength and power characteristics between power lifters, Olympic lifters, and sprinters. *J. Strength Cond. Res.* **1999**, *13*, 58–66.

23. Drinkwater, E.; Pyne, D.; McKenna, M. Design and Interpretation of Anthropometric and Fitness Testing of Basketball Players. *Sports Med.* **2008**, *38*, 565–578. [CrossRef] [PubMed]

24. González-Badillo, J.J.; Marques, M.C. Relationship between kinematic factors and countermovement jump height in trained track and field athletes. *J. Strength Cond. Res.* **2010**, *24*, 3443–3447. [CrossRef] [PubMed]

25. Taylor, S.J.; Taylor, K.-L. Normative data for mechanical variables during loaded and unloaded countermovement jumps. *J. Aust. Strength Cond.* **2014**, *22*, 26–32.

26. Newell, K.M.; Kugler, P.N. Search strategies and the acquisition of coordination. *Adv. Psychol.* **1989**, *61*, 85–122.

27. Coutts, A.J.; Murphy, A.J.; Dascombe, B.J. Effect of direct supervision of a strength coach on measures of muscular strength and power in young rugby league players. *J. Strength Cond. Res.* **2004**, *18*, 316–323. [PubMed]

28. Yang, J.-F.; Scholz, J. Learning a throwing task is associated with differential changes in the use of motor abundance. *Exp. Brain Res.* **2005**, *163*, 137–158. [CrossRef] [PubMed]

29. Kraemer, W.; Newton, R. Training for muscular power. *Phys. Med. Rehabilit. Clin. N. Am.* **2000**, *11*, 341–368.

30. Müller, H.; Sternad, D. Decomposition of variability in the execution of goal-oriented tasks: Three components of skill improvement. *J. Exp. Psychol. Hum. Percept. Perform.* **2004**, *30*, 212. [CrossRef] [PubMed]

31. Rodacki, A.L.; Fowler, N.E.; Bennett, S.J. Vertical jump coordination: Fatigue effects. *Med. Sci. Sports Exerc.* **2002**, *34*, 105–116. [CrossRef] [PubMed]

32. Stone, M.H.; O'Bryant, H.S.; McCoy, L.; Coglianese, R.; Lehmkuhl, M.; Schilling, B. Power and maximum strength relationships during performance of dynamic and static weighted jumps. *J. Strength Cond. Res.* **2003**, *17*, 140–147. [PubMed]

33. Kraemer, W.J.; French, D.N.; Paxton, N.J.; Häkkinen, K.; Volek, J.S.; Sebastianelli, W.J.; Putukian, M.; Newton, R.U.; Rubin, M.R.; Gómez, A.L.; et al. Changes in exercise performance and hormonal concentrations over a big ten soccer season in starters and nonstarters. *J. Strength Cond. Res.* **2004**, *18*, 121–128. [PubMed]

© 2017 by the authors. Licensee MDPI, Basel, Switzerland. This article is an open access article distributed under the terms and conditions of the Creative Commons Attribution (CC BY) license (http://creativecommons.org/licenses/by/4.0/).

MDPI

St. Alban-Anlage 66

4052 Basel

Switzerland

Tel. +41 61 683 77 34

Fax +41 61 302 89 18

www.mdpi.com

Sports Editorial Office

E-mail: sports@mdpi.com

www.mdpi.com/journal/sports

www.ingramcontent.com/pod-product-compliance
Lightning Source LLC
Chambersburg PA
CBHW051915210326
41597CB00033B/6152